Make: 86
CONTENTS

See more of Shawn Thorsson's fantastic cosplays on page 26.

26

ON THE COVER: Jen Schachter, aka Schacattack, wears her 80s glam-inspired Marie Antoinette wig made from EVA foam.

Images: Jen Schachter (Marie Antoinette cosplay) and Joshua Ellingson (Pepper's Ghost Flask)

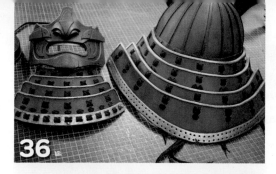

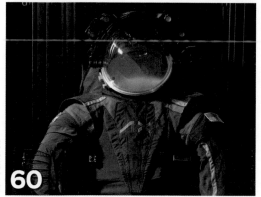

Michi, SKS Props, Axiom Space, Chad Champion, Joshua Ellingson, Arturo182, Rich Cameron

Make:®

> "If human beings had genuine courage, they'd wear their costumes every day of the year, not just on Halloween."
> —Douglas Coupland

PRESIDENT
Dale Dougherty
dale@make.co

VP, PARTNERSHIPS
Todd Sotkiewicz
todd@make.co

EDITORIAL

EDITOR-IN-CHIEF
Keith Hammond
keith@make.co

SENIOR EDITOR
Caleb Kraft
caleb@make.co

COMMUNITY EDITOR
David J. Groom
david@make.co

PRODUCTION MANAGER
Craig Couden

CONTRIBUTING EDITORS
Tim Deagan
William Gurstelle

CONTRIBUTING WRITERS
AND ARTISTS
Rhett Allain, Rehana Al-Soltane, Erin Winick Anthony, arturo182, Cabe Atwell, Jeremie Boulianne, Rich Cameron, Regan Cerato, Chad Champion, Stephanie Chan, David Dalley, Tiphani Dixon, Beverly Downen, Joshua Duart, Ben Eadie, Joshua Ellingson, Christina Ernst, Greg Gilman, Paul J. Henley, Joan Horvath, Tristian Johnson, SJ Jones, Bob Knetzger, Kelley Kullman, Audrey Love, Forrest M. Mims III, Rob Nance, Dylan O'Connell, Philip Odango, Marshall Piros, Charles Platt, Nick Poole, Anuradha Reddy, Jen Schachter, Steven K. Smith, Martin Spendiff, Lee Wilkins, Willow Creative, Sophy Wong, Lee D. Zlotoff

MAKE.CO

ENGINEERING MANAGER
Alicia Williams

WEB APPLICATION
DEVELOPER
Rio Roth-Barreiro

DESIGN

CREATIVE DIRECTOR
Juliann Brown

BOOKS

BOOKS EDITOR
Kevin Toyama
books@make.co

GLOBAL MAKER FAIRE

MANAGING DIRECTOR,
GLOBAL MAKER FAIRE
Katie D. Kunde

GLOBAL LICENSING
Jennifer Blakeslee

MARKETING

DIRECTOR OF
MARKETING
Gillian Mutti

PROGRAM COORDINATOR
Jamie Agius

OPERATIONS

ADMINISTRATIVE
MANAGER
Cathy Shanahan

ACCOUNTING MANAGER
Kelly Marshall

OPERATIONS MANAGER
& MAKER SHED
Rob Bullington

LOGISTICS
COORDINATOR
Phil Muelrath

PUBLISHED BY

MAKE COMMUNITY, LLC
Dale Dougherty

Copyright © 2023 Make Community, LLC. All rights reserved. Reproduction without permission is prohibited. Printed in the U.S. by Schumann Printers, Inc.

Comments may be sent to:
editor@makezine.com

Visit us online:
make.co

Follow us:
🐦 @make @makerfaire @makershed
📘 makemagazine
📷 makemagazine
▶ makemagazine
Ⓟ makemagazine

Manage your account online, including change of address: makezine.com/account
For telephone service call 847-559-7395 between the hours of 8am and 4:30pm CST. Fax: 847-564-9453.
Email: make@omeda.com

Make: Community

Support for the publication of *Make:* magazine is made possible in part by the members of Make: Community. Join us at make.co.

CONTRIBUTORS

If you had unlimited resources, what would be your dream character to cosplay as?

Willow Creative
Florida
(Passive Motion Masks)
I would create the Arbiter from *Halo 3*, one of my favorite creatures and close to my childhood.

Christina Ernst
Chicago, Illinois
(Candlelit Cathedral Dress)
Medusa, complete with a motorized snake wig.

Erin Winick Anthony
Houston, Texas
(Making NASA's New Moon Suits)
I've always wanted to create Elphaba's Act 2 dress from the musical *Wicked*! The costumes from the show are incredible.

Issue No. 86, Fall 2023. *Make:* (ISSN 1556-2336) is published quarterly by Make Community, LLC, in the months of February, May, Aug, and Nov. Make: Community is located at 150 Todd Road, Suite 100, Santa Rosa, CA 95407. SUBSCRIPTIONS: Send all subscription requests to *Make:*, P.O. Box 566, Lincolnshire, IL 60069 or subscribe online at makezine.com/subscribe or via phone at (866) 289-8847 (U.S. and Canada); all other countries call (818) 487-2037. Subscriptions are available for $34.99 for 1 year (4 issues) in the United States; in Canada: $43.99 USD; all other countries: $49.99 USD. Periodicals Postage Paid at San Francisco, CA, and at additional mailing offices. POSTMASTER: Send address changes to *Make:*, P.O. Box 566, Lincolnshire, IL 60069. Canada Post Publications Mail Agreement Number 41129568.

THE MAKER EFFECT FOUNDATION

- PROUDLY PRESENTS -

THE GREATEST SHOW (& TELL) ON EARTH!

MAKER FAIRE ORLANDO

NOVEMBER

SAT AND SUN | 4TH AND 5TH | 2023

FAIRE _____.COM

- PROUD SPONSORS -

FROM THE EDITOR'S DESK

MAKE: READERS BECOME MAKE: AUTHORS!

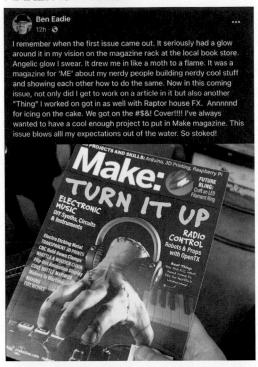

Ben Eadie
12h · ⊗

I remember when the first issue came out. It seriously had a glow around it in my vision on the magazine rack at the local book store. Angelic glow I swear. It drew me in like a moth to a flame. It was a magazine for 'ME' about my nerdy people building nerdy cool stuff and showing each other how to do the same. Now in this coming issue, not only did I get to work on an article in it but also another "Thing" I worked on got in as well with Raptor house FX. Annnnnd for icing on the cake. We got on the #$&! Cover!!!! I've always wanted to have a cool enough project to put in Make magazine. This issue blows alll my expectations out of the water. So stoked!

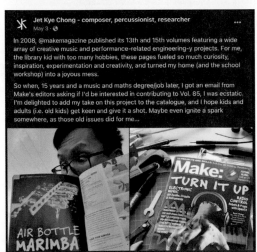

Jet Kye Chong - composer, percussionist, researcher
May 3 · ⊗

In 2008, @makemagazine published its 13th and 15th volumes featuring a wide array of creative music and performance-related engineering-y projects. For me, the library kid with too many hobbies, these pages fueled so much curiosity, inspiration, experimentation and creativity, and turned my home (and the school workshop) into a joyous mess.

So when, 15 years and a music and maths degree/job later, I got an email from Make's editors asking if I'd be interested in contributing to Vol. 85, I was ecstatic. I'm delighted to add my take on this project to the catalogue, and I hope kids and adults (i.e. old kids) get keen and give it a shot. Maybe even ignite a spark somewhere, as those old issues did for me...

GOT A PROJECT THAT BELONGS IN MAKE:?
Submit an article idea at make.co/submit-an-article-or-book-idea

UNDERSTANDING A CIRCUIT

"Pitch Perfect" (*Make:* Volume 85, page 52) is exactly my kind of project, using a few simple parts in a creative way, with no microcontroller in sight. I got quite excited as I started reading it. But I can't understand how the circuit works. Maybe more diagrams would be helpful. I desperately needed a schematic, and/or a flow diagram, so I could visualize what's happening. —*Charles Platt, author of Electronics Fun & Fundamentals column*

Author Kirk Pearson replies:
Here's a schematic of The Undertoner circuit (see full size at makezine.com/go/undertoner) that we prepared for our upcoming book, *Make: Electronic Music from Scratch.* Hopefully it can be of use!

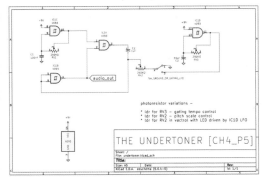

SOLAR BOTTLE POWER

Thank you for the article **"DIY Solar Bottle Lamp"** (*Make:* Volume 82, page 44). How do you connect it to a house or shed? Do you need to do any 3D printing? Is it possible to make a smaller version? —*Ben Engleman, Grade 6, Cincinnati, Ohio*

Author Debasish Dutta replies:
You can mount it to a shed by using a piece of corrugated sheet, as shown at instructables.com/How-to-build-a-SOLAR-BOTTLE-BULB. No need for 3D printing; however, you may also design a mounting arrangement by using a 3D printer. The size of the lamp is determined by the solar panel and battery. If we reduce the size, a smaller solar panel won't be sufficient for the desired output. ⊘

Make: Books

LET CURIOSITY LEAD YOUR LEARNING JOURNEY

Burn Things Out, Mess Things Up — That's How You Learn.

Make:
ELECTRONICS
THIRD EDITION
Charles Platt

Make: Electronics, Third Edition

by Charles Platt $34.99

NEW COLOR ILLUSTRATIONS!

Starting with basic concepts, this friendly and comprehensible guide takes the reader step-by-step toward circuits of increasing complexity. The principle of Learning by Discovery, pioneered by Charles Platt, uses hands-on experiments to create a lasting and entertaining learning experience.

OTHER GOODIES TO GRAB:

Make: Calculus
by Joan Horvath and Rich Cameron
$29.99

Robot Magic
by Mario Marchese
$24.99

Making Simple Robots, Second Edition
by Kathy Ceceri
$24.99

Getting Started with Arduino, Fourth Edition
by Massimo Banzi and Michael Shiloh
$19.99

Getting Started with Raspberry Pi, Fourth Edition
by Shawn Wallace, Matt Richardson, and Wolfram Donat
$24.99

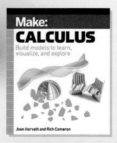

Make: Books are also available on amazon.com

Schmücken Everything You Make

by Dale Dougherty, President of Make: Community

Above the busy build area for Nerdy Derby was a sign that read *Schmücken*, which means "decorate" in German. A young girl built a wooden car decorated with pink and blue feathers and silver pipe cleaners. Of course the car sported googly eyes, as essential as the washers that served as wheels. She showed it proudly to her father and then went to the track to race it against another car — one with an orange shark fin. Her car sped down the track and crossed the finish line first. She held it up in the air excitedly for all to see. Her car was both beautiful and fast, but I shared in the joy she felt for her creation. There were many such creative sparks at Maker Faire Vienna, held in June at a former steam engine factory outside that city.

Next, I went next to Maker Faire Prague, which is supported by Prusa Research. I visited their offices where 600 printers in a 3D printing farm are working continuously, each making parts that will be assembled into new 3D printers. Prusa's printers come in distinctive orange and black colors in sizes from Mini to XL. Some 800 people also work for Prusa in this new kind of factory.

Walking the streets of Prague, I saw a sign with a 1905 quote about the School of Decorative Arts:

Indeed, this is no mindless coercive manufactory where products are made exactly to the form, but rather a lively, energetic, and creative workshop where experienced creators and fresh and enthusiastic ones join in searching for and finding new ways.

For a century or more, the industrialized production of items with machines has seemed to undermine the creative work of groups of creators. Automation threatens to remove the human-made connection to the things we use. Does the Maker Movement represent a new way of using machines creatively to make things that also reflect our own ideas and interests? At Maker Faire Prague, the machines belonged to enthusiastic makers, and their creative projects filled four halls and attracted tens of thousands of people from all over the Czech Republic.

Next I traveled to Athens, where I visited museums with an astonishing number of artifacts from early Greek civilization. Monumental buildings were dedicated to gods; marble statues made with great skill depicted Greek gods like Athena, goddess of wisdom and war. These larger-than-life men and women, known to all Greeks from myth or history, were dressed in everyday garments, including sandals.

In museum display cases were small human or animal figurines along with functional domestic items such as ceramic bowls and vases. These objects were not in any sense plain; they usually had geometric designs or symbols. From earliest times, humans have added decoration to functional items, connecting personal life to stories or myths shared by a community.

In this issue of *Make:* we look at cosplay, which is a decorative art of its own. Cosplayers usually represent larger-than-life characters that people already know from books, video games, or movies. While costumes and action figures are available as mass-produced items, cosplayers create something that's their own, yet is instantly recognizable by others, like Jen Schachter's Marie Antoinette on the cover. Cosplayers also learn their art from each other and share ingenious solutions, such as the mechanical movements of Willow Creative's wolf mask that make its snarling mouth seem real. Sophy Wong shows how to create your own character and costume from scratch, from an existing world or a new one. Cosplay provides a context for creative expression, invoking worlds inhabited by gods, superheroes, and fairies, echoing the stories we have told each other for thousands of years. ◉

Dale Dougherty

MacGyver Challenge:
Death Valley Blowout!

By Lee D. Zlotoff and Rhett Allain

CAN YOU MACGYVER A FAST FIX FOR A FEASIBLY FATAL FLAT?

We are all MacGyvers now! Watch for the next exciting MacGyver challenge on the Make: blog (makezine.com), Mastodon (@makemagazine), Twitter (@make), and Facebook (makemagazine) and enter your solutions for a chance to be featured in these pages and win Make: *goodies!*

The Scenario

You borrowed a friend's 2000 Ford Ranger pickup truck to help you move and you're on your way to return it, driving through one of the most barren and inhospitable — not to mention hottest — places on the planet: Death Valley, California. It's August and at 9am the temperature's already over 90°F, when the truck's right rear tire blows and, by the time you pull over, is a shredded mess. Lucky for you, there's a spare and a lug wrench. But you quickly discover there's no jack! You've got no cell signal out here, so calling for help is out. And it could be hours — or days — before another car comes by. There's nothing at all in the surrounding landscape that you can use to lift the truck high enough to get the spare on.

The Challenge

In a few hours the thermometer is going to hit 120°F or more. So, using whatever you have in the truck, how are you going to replace the tire before the truck is too hot to touch and you become another sad statistic that helped give this place its name?

What You've Got And it's *all you've got:*
- 2'×2' sheet of old plywood
- 6' light duty chain
- Road flares
- Deflated yoga ball, yoga mat, bungee exercise cords
- Soccer bag with shoes, shin guards, extra clothes, folding ball rebounder, soccer ball, ball pump, running drag chute
- The terrain has light brush and mostly just dirt. There are some rocks, but the biggest is the size of a basketball.
- Garden hose (but no extra water)

TURN THE PAGE FOR SOLUTIONS

LEE ZLOTOFF is an award-winning writer, producer, and director of film and TV, including *MacGyver* (1985–1992). His new production, *MacGyver: The Musical*, casts a different audience member as Mac at each performance. macgyver.com

RHETT ALLAIN teaches physics at Southeastern Louisiana University. He was technical consultant for the *MacGyver* reboot (2016–2021) and an advisor for *MythBusters*. He blogs about physics fun at rhettallain.com.

MacGyver Challenge: Death Valley Blowout!

Our Solution

You don't have a jack, but you do have the ability to lift something. All you need is air. In fact, an inflated yoga ball should do the trick. If you place the ball under the back of the truck (in front of the back wheel) and then use the soccer pump to inflate it, as air fills the ball it will lift the truck. You can put the 2'×2' sheet of plywood between the ball and truck to protect the rubber from tearing — and spread the load over a wider contact area — but we think it would even work without that. You might think the truck would be too much weight to lift, but it should work!

Recall that pressure (P) is equal to force (F) divided by area (A). We can rearrange this to see that $F = PA$. If you have a small value for pressure but a larger area, you can get a significant force. With a pressure of just 2psi and a contact area the size of a 24"×24" sheet of plywood you can lift over 1,000 pounds! That should be good enough to get just one wheel off the ground high enough to change the tire.

Many of you had potentially workable solutions using rocks and other available elements to get the axle up off the ground and/or dig beneath the flat tire until there was sufficient clearance to change the tire. So, kudos to you all for solid

Most Plausible Solution

JOE KORDZI and **DANIEL STARR**, as they both hit upon the same pneumatic solution as we did. Pump it up!

Most Creative Solution

WARREN WITHROW. Though it might be tricky to make this work exactly as described, his notion of removing the truck's tailgate — no tools required — and using it as a ramp/lever to get the right rear wheel off the ground demonstrated excellent outside-the-box thinking by looking past just the list of available objects.

Honorable Mention

JIMMY FINKE. Though his solution didn't embrace the MacGyver 'ethic' per se, it was practical and workable: drive on the flat tire and replace the ruined rim later.

Also, a special thanks to **AL P.** for letting us know that, precisely to prevent this Challenge, the police make a point of patrolling Death Valley National Park's main roads regularly should any traveler find themselves stranded — and for informing us about the special asphalt used to resist the heat! ◷

Make:
EDUCATION
FORUM

A VIRTUAL EVENT BY AND FOR MAKER EDUCATORS

During this innovative two-day virtual event, you'll dive into exciting maker topics, connect with like-minded educators, and explore new technical skills that will transform how you teach.

SEPTEMBER 22—23, 2023

make.co/educationforum

MADE
ON EARTH

Amazing builds from around the globe

Know a project that would be perfect for Made on Earth?
Let us know: *editor@makezine.com*

INTRUDER ALERT

RUNEGUNERIUSSEN.NO

Rune Guneriussen has been shining a light on mankind's intrusion on the natural world for the last 20 years, juxtaposing lamps and other human technology with organic beauty that is all too rapidly disappearing. "We are in this silly situation of actually eradicating nature for people to get closer to nature," he says while speaking to *Make:* about his work.

The Norwegian artist combines multiple disciplines, including naturalism, sculpting, photography, and miles of hiking, to produce stunning images like *A raven wails across the lowlands* (shown at left), featuring dozens of lamps fruiting like mushrooms, creating both a sense of wonder and plunder as man and nature collide.

"Mushrooms are parasites taking energy from other plants and trees. And I think that is what we are," he says while discussing the piece he shot 10 years ago, utilizing lamps powered by generators and batteries. The photo is part of an extensive series called *A parasitic gesture*, for which he also lugged telephones, chairs, and books deep into isolated ecosystems to design these thought-provoking contrasts in complete solitude.

"When you put in the positivity of a gesture, we could be something good," he says, but admits he's lost that optimism as mankind increasingly favors consumption over conservation. "Looking at the installation itself, it scares me more than it's beautiful."

He hopes his work offers audiences a closer connection to nature, which he has enjoyed since wandering the Norwegian countryside as a boy. But now, "time is running out," he says, "and the distance between us and nature has just become stronger."

Only time will tell if humankind can make an intellectual leap toward conservation, but Guneriussen is adamant, "We can't stop trying. If we stop trying, we die." —*Greg Gilman*

Rune Guneriussen

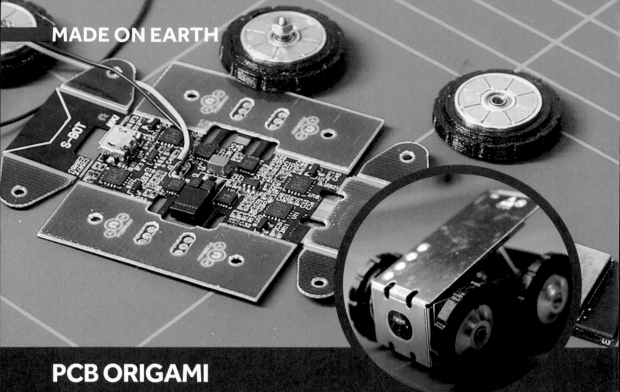

Carl Bugeja

PCB ORIGAMI

YOUTU.BE/PSZSZV4QHU4

Printed circuit boards (PCBs) are commonly thought of as a rigid, nonflexible plane that supports and connects electronic components. But today, flexible circuits are just as commonplace. The idea of a flexible circuit has been around for decades, and engineers have successfully developed integrated circuits (ICs) that can bend, stretch, and twist to a certain degree.

Carl Bugeja, an engineer from Multa, Italy, has capitalized on that flexible circuit technology and produced a number of projects that take advantage of its unique applications.

His latest project, a foldable-PCB robot rover, is unique in that it doesn't use a typical chassis but instead uses the PCB itself as the robot's outer shell. Instead of using connectors or solderable joints to create the necessary angles, it uses a semi-rigid PCB that can fold into a box shape and unfold for troubleshooting when needed. To that end, the rover is held together by bolts, which also act as axles for its wheels. Those wheels are actuated using a flat stator design Bugeja developed, which is also incorporated into the PCB and can be controlled wirelessly using a PlayStation controller.

Bugeja's PCB motor actuators are key components of the foldable rover and are designed using a 12-layer spiral PCB coil. When current flows through the internal copper windings, it generates a magnetic field that can repel or attract N52 neodymium magnets that act as actuators. That magnetic strength, up to 12mT (millitesla), is enough force to actuate lightweight objects. "The way they work is very much like a normal brushless motor. My project is a little bit different because the motor stator is embedded into the PCB, and the magnets are inserted in the wheels," states Bugeja.

Driven by a twin-cell lithium battery, the rover is surprisingly capable of traversing flat and uneven terrain using 3D-printed tires with ridges. An onboard ESP32 microcontroller handles Bluetooth communication, while four separate PIC-based driver chips commutate the motors. As with any project, Bugeja encountered several design flaws and bugs during the rover's development, and made several adjustments and even a new PCB revision to get the robot to function closer to what he imagined. He puts his ethos concisely: "Pushing the boundaries of electronics is my goal."

—Cabe Atwell

AMPHIBIOUS ABODES

INSTAGRAM.COM/UNKNOWNDAZZA

UnknownDazza

The art of making is all about creating something more than the sum of its parts, and often ideas are made possible by multiple makers working together. Some projects involve just a small cohort, while others have dozens operating in concert — but is there an upper limit to this collaborative crowdsourcing? Australian maker **UnknownDazza** doesn't think so, as his 3D-printed frog houses went viral and continue to be improved by the legions of frog and fabrication fans who follow him.

Dazza's first video on TikTok showcasing the project's earliest version dropped in October 2022 and garnered almost 3 million views, with support for the frogs (all lovingly named "Frod" by the community) continuing to increase daily. "I think, in general, people are interested in things that take a lot of effort," he said in an interview. "The less effort something is, I think the less remarkable it is, so the fact that six months of hundreds and hundreds of hours of work compressed into, like, two minutes, got so much support ... it means a lot." While he acknowledges that most of his general audience is subscribed just to see the Frods and their possum neighbors, nearly every update video receives dozens of suggestions from commenters on design improvements. Dazza, who works in software as a product designer and manager, loves the crowdsourcing element and takes every opportunity to shape the Frod House according to "the will of the people." Elaborating, he states, "Having this resource, this massive pool of ideas and then being able to take the best ones ... I think that's quite a cool process."

Because of commenters' suggestions, the Frod House now features an open-air porch and windows, a pool with an infinity waterfall fountain, an emergency escape cave in case of danger, and a tadpole hatchery with a motorized ramp. Dazza's favorite additions are the pool, for its versatility and usefulness, and the tadpole ramp, since he loves adding complex mechanisms to his projects and he enjoyed 3D printing the ramp's chain as a single linked entity. In fact, he hopes to continue crowdsourcing future projects while adding mechanical elements.

What's next? "I'm almost done with a much larger house for the possums that combines wood and 3D-printed materials," says Dazza. "I'd also always imagined adding a feeding station that could open up and release food for the Frods, and I thought that could be like a cool live activity on the livestream."

—*Marshall Piros*

Lego-based PancakeBot at World Maker Faire in New York, 2012.

Papa's Gonna Build You a

PancakeBot

How food printing became a family affair

Written by Dale Dougherty

Miguel Valenzuela had just finished dinner when we started talking over Zoom. "I'm living in Norway. My sister-in-law came over with her family from Denmark. Typically when they come over, I usually make some Mexican food for them and we just found these awesome tortillas." He explained that on a trip to Barcelona he met a Mexican man who makes nixtamal tortillas in the traditional way. He continues to buy them and share them with family and friends in Norway. "I've been struggling, especially being from Southern California and Northern Mexico,

missing my food. I've been trying to make a home of it here."

Miguel and his future wife, Runi Elisabeth Flata Steen who is from Mjøndalen, Norway, met at Cal Poly San Luis Obispo in California while he was finishing his engineering degree. Thirteen years ago, they moved to Oslo. "We were living in a small little apartment," said Miguel who then had two young daughters, Lily and Maia. "I had all these Legos with me that my friend Mark Dumas had given to me. I was just trying to figure out how to stay sane because I went from a lovely San

Diego sun to a meter of snow over here."

He read the second issue of *Make:* magazine in which author Bob Parks in the article "Blockheads" called Lego the "ultimate prototyping tool." Parks mentioned a person in the UK who had made a pancake-stamping machine out of Lego. As Miguel mentioned the project to his daughter Lily, she turned to her sister, Maia, and said: "Papa's gonna build a pancake machine out of Legos."

PROTOTYPING IN LEGO

This initial project went through 13 different iterations over 6 months. Miguel's goal was to predictably make two Mickey Mouse pancakes. That's three blobs of batter on a griddle. The entire drawing machine was made of Legos, except the ketchup bottle that held the batter. He used Lego NXT, which could control three motors. Two of the motors controlled motion along the X and Y axes. "I didn't know how to program very well, so I had to do everything by hand," he said. He drew a figure on graph paper, and then input the degrees of rotation positive or negative into an Excel spreadsheet. Then he'd import it as a text file that his program read with three lines per movement. The third motor controlled the flow of the batter. "I set it up to have a strange little valve structure. I had to pump air and control the valves all with one motor, and I did it by reversing the direction of the motor."

His canvas was a Presto griddle, about 17 by 8 inches (45 by 22 centimeters). "The squeeze bottle would go back and forth while dispensing the batter to draw, which was the equivalent of pen-up and pen-down commands. Pen up reversed the air flow and pen down added pressure to dispense it." A lot depended on getting the viscosity of the batter right. "I think I went through 20 gallons of pancake batter trying to figure this thing out in my kitchen at home."

He published the instructions for building a PancakeBot out of Lego on SourceForge. They are still available. "I remember telling my wife, I'm gonna see if we can put this out there." Before releasing the instructions, he got the domain name, pancakebot.com, as well as other social media handles. "You gotta stake your claim on the internet," he said, adding that he did it just in case

Miguel Valenzuela, creator of PancakeBot.

"I think I went through 20 gallons of pancake batter trying to figure this thing out in my kitchen at home."

it turned out to be worth something.

Initially, PancakeBot was just for fun; Miguel would later call it an "unnecessary invention." What excited him was not just that it worked but that PancakeBot made kids smile. "My daughters were cracking up," he said. It didn't matter what the pancakes looked like. How much fun is a machine that can draw pancakes that you can eat? That came across in an early YouTube video he made with his two oldest daughters in 2011. Maia presses the button on the NXT controller and soon a Mickey Mouse pancake is printing out.

Word started to spread. Inventor/entrepreneur Jeri Ellsworth tweeted about his videos. Goli Mohammadi of *Make:* wrote about it. "We had, all of a sudden, all this attention for this crazy machine built from Legos."

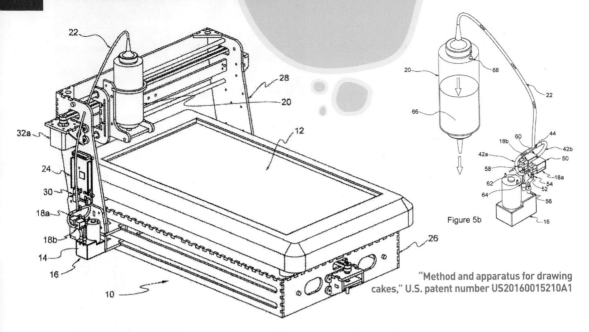

Figure 5b

"Method and apparatus for drawing cakes," U.S. patent number US20160015210A1

FROM MAKER FAIRE TO THE PATENT OFFICE

Miguel and Runi brought the PancakeBot to the New York Maker Faire in 2012. "That was an amazing experience. The attention that this thing got from all the kids — they're standing around waiting 15 to 20 minutes for little round pancakes. I guess it was the combination of pancakes, Legos, and robots." He brought the whole family to the first Oslo Maker Faire in 2013 and he was featured on Norwegian TV, which pleased his mother-in-law.

Yet it started to bother him that PancakeBot was not taken more seriously. "I didn't know if there was anything serious about it," he said, thinking that maybe it was a kind of product that others might buy. In 2014, he recalls asking Runi: "What are we going to do with it? Should we give up on it?" Runi, who is an actress involved in children's theater and a middle school teacher, remained supportive. She said to give it another try and build a new prototype.

Miguel began using components that had become more available to makers and hobbyists. He also made connections in the maker community that helped him take PancakeBot to the next level. He found new motor shields from Adafruit. Dan Royer from marginallyclever.com had code for the motor shields. He started talking

to Windell Oskay of Evil Mad Science Laboratories who had an Eggbot product that inspired him. Soon, his new prototype had a chassis made out of laser-cut acrylic and a redesigned pneumatic system.

In 2014, he applied to Maker Faire Bay Area and was accepted even before the prototype was really finished. Fortunately, he completed the new prototype just in time to board the plane. Unfortunately, when he unpacked the PancakeBot after the flight, he discovered that the acrylic chassis had broken.

He reached out to Oskay. "Evil Mad Scientist stepped in and they helped me laser cut the new pieces." The response at the Bay Area Maker Faire in San Mateo was "phenomenal and overwhelming." He thought it "validated the whole concept of food printing in general."

In 2015, he got a patent on a PancakeBot: "Method and apparatus for drawing cakes," US patent number US20160015210A1. "The way it controls the fluid pneumatically without having any mechanical elements inside of it was unique," he explained. "I wanted to just show that you don't need a plunger, like on a syringe," he said.

The PancakeBot had become a little more serious for Miguel, but it was still a side project, work he did outside of his job as a design engineer.

Miguel at the White House Maker Faire in 2014.

PANCAKEBOT BECOMES A PRODUCT

Maker Faire provided a market validation for PancakeBot. People were asking Miguel when he was going to do a Kickstarter and create a product that they could buy. He recalls other makers talking about creating a Kickstarter — that was what you did with a successful prototype. Yet he never got around to doing his own Kickstarter, even though he kept thinking about it.

One day, he got a call from a new company that sought to help inventors bring products to market. The company's CEO asked Miguel if he'd be willing to license the PancakeBot to him. "Living in Norway and knowing the market for the product was the United States, I thought this was probably the right thing to do," said Miguel. While thinking over that decision, he got an email inviting him to participate in the White House Maker Faire in 2014.

"I actually just had to sit on the couch for a second because here's this crazy machine that started off with Legos and a whimsical idea and now we're getting invited to go to the White House," he recalled. The invitation said that there was only room for Miguel. "I can't really go by myself because this is a family thing. My wife was

> "I actually just had to sit on the couch for a second because here's this crazy machine that started off with Legos and a whimsical idea and now we're getting invited to go to the White House."

supporting me and she was at the Faires with me. The girls were there flipping pancakes, wearing PancakeBot T-shirts, and Lily had inspired it. I would've felt really bad if I would've gone by myself." The organizers extended the invitation to his whole family and they all went to Washington to make pancakes for President Obama. Miguel, his wife, and his daughters were dressed as chefs for the White House event.

"That opened up even more doors for us and got us a lot of attention," recalled Miguel.

Miguel with the Lego-based PancakeBot and the later acrylic model at Maker Faire Paris, 2014.

The design above will resemble the actual pancake once it has been flipped.

PancakePainter software for drawing images for PancakeBot.

However, the success also caused a lot of stress. "If you're not ready to take your product to market and you get all this attention, then either somebody's gonna steal it or copy it," he said, worried that he might miss the opportunity to have PancakeBot become a product.

In early 2015, Miguel agreed to license the product and turn over development to the new company. Work began on a new 1.0 version made in China. And finally, PancakeBot did a Kickstarter. "We were able to meet our goal of $50,000 within the first 24 hours," said Miguel. On a visit to Kickstarter, Miguel learned that a lot of people had been asking Kickstarter where to find the PancakeBot. They told him he had lots of fans, and he was overcome with emotion.

The Kickstarter raised more than $460,000 from 2,074 backers. Miguel, Runi, Lily, and Maia appeared in the video explaining PancakeBot to potential backers, showing how to take a drawing of a rocket ship, scan the drawing and produce code that PancakeBot uses to draw with batter. PancakeBot 1.0 was delivered about 8 months after its Kickstarter was over. "The first run was 3,500 units, and 2,000 units were going to Kickstarter backers," he explained.

> "The PancakeBot is just a small part of a bigger way of learning, but it keeps the kids interested. You can have such a quick result and then if you're doing pancakes, you can eat it."

As excited as Miguel was that he had finally had a real product in the hands of consumers, he learned that these consumers had a problem with PancakeBot 1.0. The original prototype had a potentiometer that allowed you to adjust the speed based on the viscosity of the batter. This was dropped from PancakeBot 1.0. "The engineers locked in the speed of the PancakeBot," said Miguel, which meant that speed adjustments couldn't be made — PancakeBot was very slow and frustrating. Miguel had to scramble to distribute a firmware update along with detailed instructions to all the Kickstarter backers, some of whom were frustrated at having to go through this process. "I think that took us back a little bit," said Miguel. He also learned about other problems that users had with the machine, such as overfilling the bottle with batter and having it clog the tubes.

By 2019, a new and improved PancakeBot 2.0 had come out and it sold at retail for between $250 and $350. "We were in Bed, Bath & Beyond, Best Buy, and all these fancy stores, but a lot of them ended up in schools and in makerspaces," said Miguel. Norwegian engineer Jan Dyre Bjerknes developed PancakePainter, a free drawing program written in Processing, that allows anyone to create new drawings and save them as G-code for the PancakeBot (github.com/pancakebot).

Miguel noted with interest that PancakeBot was being adapted for other uses. "A company in Houston was using it to dispense glue on CCD screens," he said, adding that he helped them with their G-code.

STARTING OVER AGAIN

In 2019, Miguel and the company that licensed PancakeBot parted ways. "In the end, they did not see or understand my vision. PancakeBot was not an appliance, it was the Easy Bake Oven of the 3D food printing world," said Miguel. He got the rights back and took on some leftover inventory, which he's been moving with the help of the Fab Foundation.

The basic machine works well but he thinks there could be variants for different markets. One is education, where it is used by students and teachers in the classroom. He's now at work developing curriculum for PancakeBot. "That is really organizing activities around PancakeBot, showing what it can do," he said. He has worked with a teacher who likes to have the kids make pancakes with geometric shapes. "The children talk about why they like the shape, how it was made," he said. "The PancakeBot is just a small part of a bigger way of learning, but it keeps the kids interested. You can have such a quick result and then if you're doing pancakes, you can eat it."

The PancakeBot is still growing up, just like Lily who is now 15. Perhaps she is taking on the role of a product manager vetting new features. "PancakeBot's a child of the Maker Movement so that's why we have to keep supporting it," said Miguel, speaking of this whimsical machine that became a family's pet project and more. ◔

DALE DOUGHERTY is the founder of *Make:* and president of Make: Community.

Drawing by Lily of the family.

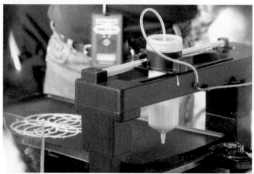

Pancakebot 1.0 in action at Maker Faire Bay Area, 2014.

Miguel and his two daughters with the PancakeBot.

Quentin Chevrier, Miguell Valenzuela, Lenore Edman, Lily Valenzuela

PancakeBot product site:
pancakebot.com

PancakePainter software:
github.com/pancakebot/pancakepainter

DIY instructions for Lego version:
sourceforge.net/projects/pancakebot

Video of Lego version:
youtu.be/2aux0ZQJVBk

DISSECTING THE DIY DIODE

Building a functional vacuum tube in the home lab isn't easy, but it's not impossible!

Written and photographed by Nick Poole

NICK POOLE is an electronics engineer, product designer, and serial hobbyist in Southwest Virginia. A long-tenured team member at SparkFun Electronics, he spends his free time doing other nerdy things that you can follow on his YouTube channel SignalDitch.

A BIT OF INSPIRATION

In 2022, I was skimming through YouTube videos to pass the time when I stumbled across a channel called Usagi Electric. The channel belongs to a guy named David who — among other things — is building a vacuum tube computer with a 1-bit processor. The processor alone is about the size of a coffee table, and comprises over 200 6AU6 pentode tubes. It's a work of art.

That said, I couldn't help but think about making it smaller. The miniature vacuum tubes in David's processor are already small, about ¾" in diameter and 1½" tall. But even with tiny tubes, a computer built from discrete switching units can only get so small. If we want to really pack in the bits, we need to look at what happened with transistorized computers. Once transistors got small enough, we stopped using discrete transistors and we started to package them together into logic units. We created the integrated circuit.

THE COMPANY THAT NEVER WAS

Imagine with me that the transistor was never discovered, but miniaturization kept going. Instead of integrated circuits made from patterning and doping silicon oxides, maybe we would be building microscale pentode arrays on sheet nickel inside of evacuated leadframes.

I started to explore this idea by making 3D renderings of what these imaginary devices might look like. At the time, I was doing a lot of CAD for my day job and I was looking for a project to level-up my modeling skills without having to worry about the practicalities of mechanical design. I started to post my renders online, and people got excited about them, so I turned it into a sort of art project. I invented a fictional company called Integrated Thermionic and assigned part numbers and branding to my creations. I went so far as to design packaging and sales materials for Integrated Thermionic, with the eventual goal of publishing an art book in the style of a catalog.

Every now and then, a particularly convincing rendering would escape its context and send folks to eBay looking for a tube that never existed. At that point, I started to seriously consider just how difficult it might be to make my own real vacuum tubes. Of course I could make little sculptures out of resin that resembled vacuum tubes, but that didn't feel satisfying in the same way. Thus began one heck of a learning experience.

A 3D CAD rendering of Integrated Thermionics' imaginary product offering, complete with packaging.

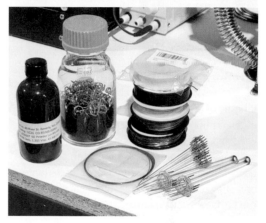

Materials such as evaporable getters, cathode coatings, and tube stems are often hard to procure.

TUBES 101

To understand how vacuum tubes are constructed, you have to understand how they work. Like semiconductor diodes, they're a sort of check-valve for electricity, allowing current to flow in only one direction. The diode owes its name to the fact that it has two (*di-*) electrodes. But in a vacuum diode, the two electrodes don't touch each other, so how does any current flow at all?

The answer lies in the phenomenon known as *thermionic emission*. The free electrons in the metal of the cathode normally can't escape the surface, but with enough thermal energy they can overcome what's called the **work function** of the metal and make the leap into free space. We can make this happen by passing a current through the cathode, like a lightbulb filament. If the other electrode (the anode) has a positive charge relative to the cathode, the field between the electrodes will accelerate these electrons into the anode, resulting in current flow. The trick is that the anode is *not* hot, so even if we reversed the voltage across the tube, the electrons on the anode wouldn't have enough energy to jump across to the cathode; this results in the one-way diode behavior that we've come to expect.

This is the principle behind all *hot cathode* vacuum tubes. Once you understand the diode, it becomes easier to understand all other types of hot cathode tubes. A triode, for instance, adds an electrode between the anode and the cathode, called a **grid**. By controlling the voltage between this grid and the cathode, varying amounts of current can be allowed to flow through the tube, resulting in a kind of switch, or amplifier. Keep adding elements to create tetrodes, pentodes, hexodes, heptodes, and so on.

The *vacuum* in a vacuum tube is necessary to keep gas molecules out of the way of the electrons. In air at standard temperature and pressure, an electron can travel about half a micrometer before it hits a gas molecule. Inside a vacuum tube, with a pressure of about a billionth of an atmosphere, the free path of an electron is several times any reasonable path inside the tube. This way, electrons aren't likely to smack into any gas molecules, ionizing them and causing other adverse reactions.

BUILDING A DIY VACUUM TUBE

The best way to explain the process of building your own vacuum tube is to walk through an example, so let's build a diode step-by-step. This is the second such device that I've attempted to build, so it's still rough around the edges. But even so, it should demonstrate the process nicely.

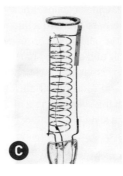
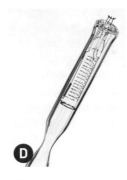

1. MAKE THE ELECTRODES

Electrodes were prepared by spot welding short lengths of fine tungsten wire onto nickel wire supports (Figure **A**). The electrode assembly was cleaned thoroughly with a mix of nonpolar solvents. After cleaning, the tungsten portion of each electrode was brushed with a torch flame and allowed to cool from a white heat. This forms the surface oxide which wets to the glass and allows us to form a vacuum seal. Any loose oxide is rubbed off using a cotton swab.

2. WELD ELECTRODES INTO GLASS

I prepared a piece of common "33 expansion" borosilicate glass tubing by heating it to softness and using a pair of pliers to gently flatten one end so that it was almost closed. I placed the electrodes into the flattened section of glass and then heated and gently smashed it on either side of the tungsten section. To finalize the seal, I heated the glass to as soft as I could using my MAPP gas torch and gave it a good squeeze with the pliers (Figure **B**).

3. ADD ANODE, FILAMENT, AND RING

The rest of the internal tube elements were formed and welded into place. The anode is a coil of nickel wire, made by wrapping it around a tool handle. The filament is 0.01mm tungsten wire and is attached to the nickel uprights simply by crimping each end. A barium ring getter was welded to the top of the anode (Figure **C**). This getter was a commercial sample from a manufacturer in China and can be difficult for hobbyists to acquire, but a ring of titanium wire can be used to similar effect (see Step 7).

4. WELD IT ALL INTO A TUBE

The entire assembly is nested inside of another glass tube (Figure **D**). While slowly turning the entire assembly, the overlap of the glass tubes was heated until soft and sealed together. The other end of the tube was heated and pulled to create a thin "neck" in the tube which would assist in sealing. The entire assembly was then attached to a vacuum compression fitting.

5. EVACUATE THE TUBE

The vacuum compression fitting was attached to the evacuation bench. This is an apparatus that I built from various used and surplus components, but it mostly boils down to two vacuum pumps: a Welch 1402 rotary-vane pump and an Edwards

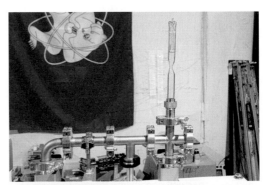
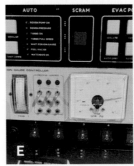

EXT255H turbomolecular pump. The system was pumped down until the ion gauge indicated 10^{-5} millibar. I heated the tube with a propane torch to drive out as much water as possible and then used a benchtop power supply to heat the filament for several minutes to drive out oxygen in the wire (Figure **E**).

6. SEAL THE TUBE

The neck of the tube was heated with a MAPP torch and carefully allowed to collapse in, sealing the tube (Figure **F**). Concentrated heat and a twisting motion ensured that the end of the tube was securely sealed.

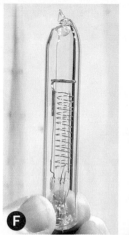

7. ACTIVATE THE GETTER

The getter that was installed earlier is activated by heating with an induction heater (Figure **G**). The getter contains reactive metal that, when heated, evaporates onto the inside of the glass and reacts with any remaining oxygen molecules to sequester them. This is the silver or black coating you see on commercial vacuum tubes. The tube is complete.

WHERE TUBE NEXT?

Currently, I'm in the process of improving my toolchain (Figure **H**). The evacuation bench has been rebuilt and is getting a custom controller to automate some of the protocol. I'm also developing an open-source glassblowing lathe specifically designed for tube building.

Once the toolchain is settled, there are a few projects on the horizon. First, I want to tackle vacuum fluorescent displays (VFDs) — I already have some cathodoluminescent phosphor set aside for that project. Second, I'd like to re-create some of my sillier Integrated Thermionics "products" such as the micro x-ray bulb or the surface-mount triode. I also have some ideas about highly integrated low-voltage thermionic logic devices that I'd like to develop.

Most importantly, though, I want to help other people get into the hobby. Vacuum tubes have been around for over a century, and while they may be obsolete in most applications, I think there's still a lot left to explore. ●

MAKE: BELIEVE

These are just a few of the awesome cosplays created by Shawn Thorsson, author of *Make: Props and Costume Armor*. Watch for Volume 2 in 2024!
makershed.com

Make:
Props and Costume Armor
Create Realistic Science Fiction and Fantasy Weapons, Armor, and Accessories

Shawn Thorsson

CELEBRATING
COSPLAY,
HALLOWEEN, AND
FANTASY FASHION
— GATEWAYS TO
CREATIVE MAKING

Written by Keith Hammond

I was a 10-year-old *Star Wars* geek in 1977, even before I saw the movie. By the time I talked my parents into driving 3 hours to Salt Lake City to see it, I already loved its heroes and villains, its fantastic ships and robots and monsters and planets. I had collected a scrapbook of news clippings and a shoebox of bubble gum trading cards. As a restless kid in a desert farm town, I knew Luke's galaxy was the world for me.

The movie was even better than I'd hoped. The FX! The creatures and costumes! The epic struggle of good and evil! And wow — the filmmakers' utter commitment to creating a rich new world for our imaginations to inhabit. I felt a part of something wonderful. I felt admiration and gratitude. Still do.

So as a fanboy, I admire cosplay as a creative expression and a community for sci-fi, fantasy, anime, and video game fandoms. What better way to celebrate fantasy worlds than to inhabit them, role-playing as your favorite characters? But I also love how cosplay and Halloween have become gateways to all kinds of making. What starts with some sewing, cardboard, and paint, soon leads to EVA foamsmithing, 3D printing armor, wiring LEDs, and designing your own fantastic fashions. In this special section, we'll show you how.

Today my 10-year-old daughter's room is full of Marvel posters and superhero masks. When I took her to our local comic-con, we saw a cosplay contest, met famed fabricator Shawn Thorsson and charity costumers from the *Star Wars* 501st Legion, and posed for selfies with Iron Man. In the cosplay universe, she can revel in the FX, the creatures and costumes — and imagine her own place in the epic struggle of good and evil. ◑

Michi, Adobe Stock-Arlenta Apostrophe

DESIGN AN ORIGINAL COSPLAY

AN EXPERT PLAYBOOK FOR BRINGING YOUR OWN FANTASY CHARACTERS TO LIFE

Written by Sophy Wong

SOPHY WONG is a designer, costumer, and maker creating artistic wearable technology. Using digital fabrication techniques like 3D printing and laser cutting, her work highlights the intersection of technology and design for the human body.

Kim Pimmel, Sophy Wong

As cosplayers, we're often focused on bringing characters from our favorite fantasy worlds to life — but have you ever wanted to make a completely original costume based on a character you developed yourself? Whether you want to create a new character within an existing fantasy world, or invent someone from an entirely new universe of your own creation, designing an original character is a fun way to tell a new story and make a costume that is one-of-a-kind. Let's take a look at some tips and strategies for making your design compelling and unique.

WORLD BUILDING AND BACKSTORY

Every character inhabits a *world.* If you're creating a new character based on an existing world, maybe from your favorite book, movie, or anime story, then you'll want to study that world and use recognizable elements from it in your design. Look closely at the details that are described or depicted in the story, and make a list of things that feel like they're unique to that world. When designing your costume, include a few specific visual elements from the world your character belongs in. This could be a badge or insignia, a pattern or texture, or even a specific color palette.

If you're creating your own fantasy world, you'll need to decide what elements inhabit it. What is the overall mood of the world? What mechanics affect the people in that world? What are the materials that exist there? In an enchanted forest full of magic and mythical creatures, the materials could be natural wood, leather, and perhaps an invented metal with special magical properties. In a futuristic sci-fi metropolis, the materials could be more engineered, like carbon fiber, titanium, and glowing neon.

In addition to the world itself, spend some time thinking about the identity of your character. Start with general things, like what time period they live in, and work your way in to details like age, gender, language, name, favorite food ... you can get as specific as you like! You might not use every detail you generate, but the goal is to establish an *identity* for the character that can inform their outfit. Here are some questions to ask yourself to help generate some ideas about who your character is:

This spacesuit costume was designed with a specific character and world in mind, for an original sci-fi short film.

- **What does this character do?** And why? This might be an occupation, quest, or goal.
- **Where is this character?** Where do they live? Where are they from? Where are they going? How did they get there?
- **When does this character exist?** In the present? Future or past? Parallel universe?

Answering these questions will help you write a *backstory* to explain the origin of your character. This is an incredibly useful exercise to do before beginning to design. You may find that while writing the character's backstory, visual ideas for the costume just come to you — keep a sketchbook handy so you can catch those great design ideas! Later on, if you get stuck on working out details while designing or building, you can read through your backstory again to find design ideas that make sense for the character.

Understanding the world and backstory of your character are the keys to developing a well-designed original costume. Ideally, you'll have all this figured out before you start designing, so that all your design decisions can help tell this story. In reality though, you may be so excited to build your costume that you want to blitz through the character development phase. That's fine too! As long as you have a general idea of who your character is, you can work out the details as you go and adjust your design to fit.

CASE STUDY: AN ORIGINAL SPACESUIT DESIGN

I recently completed an original character costume myself: a spacesuit for a filmmaker's short science fiction film. This means the world building and backstory were already developed, and I designed the costume to fit. I was given a story outline and description of the world that will be depicted in the film, as well as information about the character's identity and story arc. I used all of that, plus related research I did on my own, to come up with the design for the suit. Although I was working with an established world and character identity, I still had lots of room to get creative with the design and build process. Having so much information on the world and story made it easy for my imagination to come up with ideas for an original design.

RESEARCH AND INSPIRATION

Once you have your world and backstory figured out, it's a good idea to do some research and gather images for inspiration. With an original character, you'll need to invent every detail on the costume, so it's helpful to have reference materials on hand while designing. For an existing fantasy world, your research might involve rewatching the film and taking notes or sketching out the details you want to borrow for your design. For a fantasy world of your own creation, you may want to borrow elements from other fantasy worlds and blend them together.

Collect as many images as you can to depict the world and the people who would inhabit it. In your initial search for inspiration, gather more images than you think you need. Then, do an editing pass: lay all the images out together and remove any that aren't quite perfect for the character you are creating. Aim for 10–15 images that you could combine to describe the costume you'll be making. Pin these to an *inspiration board* so you can refer to them while designing (Figure **A**).

FANTASY SPACESUIT INSPIRATION

For my original spacesuit design, I looked at costumes from famous science fiction movies and historical flight suits and spacesuits. On my inspiration board, I omitted any "spacewalk" type suits that would be used for zero-gravity environments — instead, I focused on spacesuits used for planetary exploration. The setting for the film is a natural, Earth-like planet and the

character has an element of mystery in the story. The director of the film wanted the character to feel like an underdog within the world, so I decided to make the suit look bulky and heavy, instead of form-fitting and ready for action. I referenced old diving suits and anime mecha character designs for ideas on how to bulk out the suit with details that would look functional.

DESIGNING YOUR COSTUME: WHAT GOES WHERE?

Before you cut into any fabric or glue any foam, make a visual plan for what you're building. Sketch, collage, or trace bits of your inspiration to create a *design drawing* of the costume you will create (Figures **B** and **C**). The goal here is to work out your design ideas on paper so you have a strong plan going into your build. Don't worry about making museum-quality drawings, think of your sketches as design tools. To make sketching faster and easier, start with a *croquis* — a printed outline of a human figure — and draw your costume ideas over it. You can also take a picture of yourself and draw right on top of that!

Again, start with general ideas and work inward to details. Draw the outer shapes of your costume first; whether it's a long and flowing garment, or a fitted superhero suit, try to outline the silhouette. Then use your backstory and research for ideas to fill in the rest. Mash up lots of different details and concepts from your inspiration to make something new. Refine and evolve your design over a few rounds of drawing and redrawing to give it some distance from the inspiration sources and make it your own (Figure **D**).

While drawing out your creative ideas, also think about practical aspects of your costume, like how the pieces of the costume will anchor onto the body (usually around the waist, torso, arms, and legs). Try to visualize how gravity will make soft elements like fabric hang over the body (from the shoulders downward, and from the waist to the floor). You won't be able to predict every practical requirement of your costume at this stage, but try to at least plan for how to get in and out of your costume easily. If your costume will have electronics in it, think about where the batteries could go, and how you will turn it on and off.

Don't plan on getting your design right in the

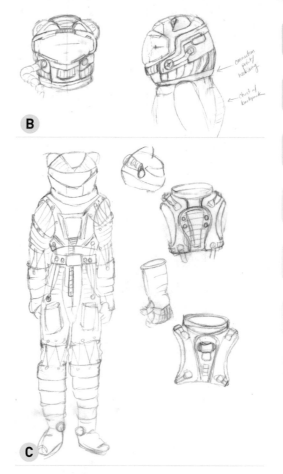

B

C

D

Sophy Wong

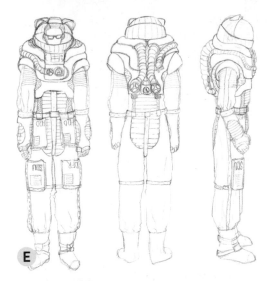

E

F

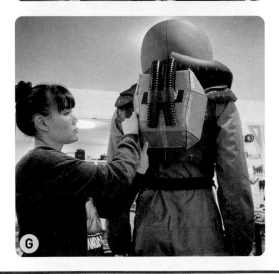

G

first sketch. Even if you think your first drawing is perfect, draw several more versions of your idea and try to make each one different. Keep drawing things until you get something you really like. Then draw your favorite design from different angles — front, side, and back (Figure E). This will force you to think through your ideas and have a more realistic plan before you start building. As always, plans can change along the way, and since you're the designer, it's up to you to decide what's right for your project.

BUILDING YOUR COSTUME: BRING IT TO LIFE

With your design sketches in hand, you're ready to start your build! This is the exciting part, where you get to use materials like foam, fabric, and paint to make your character real. Your design sketch can now become a to-do list, a diagram for all the parts you'll need to make to complete your costume. For example, from my spacesuit sketch I could see that the costume would consist of seven parts: helmet, chest piece, jumpsuit, gloves, boots, harness, and blaster. I made rough prototypes for each piece first (Figures F and G), and then worked through the build one piece at a time, always referencing my design drawings to make sure all the pieces would work together.

When choosing materials for your costume, think about the world your character lives in. What materials exist in that world? And how can you fake those materials with something more practical and wearable? Many beautiful cosplays are built with simple materials like EVA foam, cardboard, and plastic, and are painted to look like metal, feathers, leather, and more. The spacesuit I built is made almost entirely out of EVA foam, fabric, and plastic (Figures H through Q). The result is a lightweight suit that's fairly comfortable to wear and can be packed for shipping and travel.

One design technique that can help make lighter materials like foam and cardboard look like metal is to add raised details to your surfaces (Figures R , S , and T). To make something look mechanical, you could add screw heads here and there, or maybe a rectangle that looks like an access panel. To make something look like it was sculpted out of metal, you could add a raised

Sophy Wong, Kim Pimmel

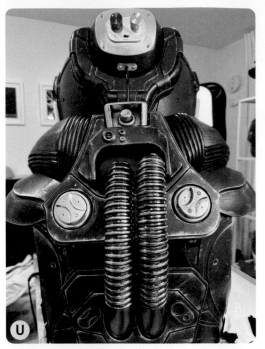

ornamental pattern or texture on top of a flat surface. When painted and weathered, the edges of those raised details can catch metallic paint and become highlights that really sell the look of worn metal. I used this effect on many of the EVA foam parts of the spacesuit. The metallic edges ensure that the surface details of the suit are visible even in low-light environments.

Another powerful design technique is to limit your *design palette*. Try to use only three different colors, or three different textures, throughout your entire costume. (Three is usually the magic number for me, but: experiment!) Using a small set of colors, textures, or motifs on the different

parts of your costume will make it feel like a unified garment (Figures **U** and **V**). Conversely, if you want your costume to look like a patchwork of scavenged parts assembled together, you may want to avoid repeating colors and motifs, to make everything feel disparate.

While working through your build, use your design sketches and research to balance details with the big picture. For each detail you want to add, ask yourself "Why is this here? What material would it be made of? Who made this?" and your answers will inform what that detail should look like (Figures **W**, **X**, and **Y**). Sometimes you'll want to hide a functional part

Z

Aa

of the costume by making it look like something else. For example, on a mage costume you might disguise a battery pack by making it look like a magical spell book. In other cases, you may be able to hide functional things in plain sight, like disguising the seams in foam armor as ornamental design lines.

You can also use your design sketches to stay on track with scale and proportion. If you drew your sketch over a croquis or a photo of yourself, then you can reference your own body measurements to help determine how big each piece should be (Figure **Z**).

Work through building each piece of your design until your costume is complete. Try on your costume pieces regularly during the build process and make any fitting adjustments you need (Figure **Aa**). A good fit is important for comfort, safety, and for ensuring that your finished project looks great.

ENJOY YOUR ORIGINAL CREATION!

After all the crafting, sewing, painting, and weathering is done, it's an amazing feeling to put on your costume and see your original design in its physical form.

Don't be surprised if your costume turned out a bit different from your design sketches; that's a normal part of solving design challenges that come up along the way. Being the designer means you have the power to sculpt the character to be anything you can imagine. And being the maker lets you bring that character out of your imagination and into reality! ◔

Sophy Wong, Kim Pimmel

STEVEN K. SMITH (aka SKS Props) is a prop and costume fabricator located in St. Louis, Missouri, with his wife and two daughters. He has made a name in the maker community by creating highly detailed builds focused on pop culture references and characters.

FOAMSMITHING
FAUX LEATHER
HOW TO SIMULATE LEATHER USING EVA FOAM

Written by Steven K. Smith

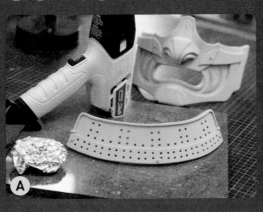

A

B

SKS Props

On my YouTube channel I build things that inspire me — sometimes that's a new movie, sometimes a video game from a decade ago (youtube.com/SKSProps). My latest project is one I've had on my list for quite some time: a full set of traditional samurai armor! Like my other projects, free patterns for this are on my site if you'd like to make one too (sksprops.com).

As a seasoned prop maker I have worked with lots of different media over the years, from clay and resin to 3D modeling and foam. **EVA foam** is by far my favorite! In the cosplay community I'm known as a "foamsmith," or one who specializes in foam fabrication. EVA is one of the best materials to build with, especially for Makers new to foamsmithing, because it is lightweight, inexpensive, and extremely versatile. Once you learn the fabrication techniques you can make foam look like skin, wood, rock, bone, and metal!

THE LEATHER LOOK

Traditionally, samurai armor was made of iron plates or strips of leather, lacquered to harden the surface and to protect them from the elements. To simulate a leather texture on the foam, I'm going to use one of my favorite tricks, the **tinfoil technique** (Figure **A**). EVA foam takes heat well and if you press into it while it's soft, that will leave the surface embossed with the applied design or texture. You can press lots of different materials and shapes into the foam to give it a unique look.

Before you start, let's talk about safety: Always wear a proper respirator and eye protection when heating or sanding foam. Lightly heat the surface of the 4mm foam with a heat gun, moving it back and forth and taking care not to overheat the foam, because it is possible to burn or melt the surface. Once the foam has softened, quickly press some crumpled-up tinfoil into the surface. If necessary, you can rotate the foil ball to give the embossed pattern a more random appearance.

Once the foam cools the applied texture will remain in the foam (Figure **B**), unless the foam is reheated, in which case it will bounce back to its original shape. So this is a finishing process that should be done at the end of your build, right before paint and primer are added. Once the foam has been sealed with a product like **Plasti Dip**,

Rosco Flexbond, **Creature Cast**, or **Mod Podge**, the textures will be locked in.

THE POWER OF PAINT

That leads us to what could be the most important part of prop making, and that is painting. I've been a builder for 10 years, but as a traditionally trained illustrator I've been painting for 30. Almost all of my projects are hand painted. After sealing the foam, I'll usually do a light dusting of a rattle-can **spray paint** to act as a base layer of color (Figure **C**). This is especially important if you're going to have a lighter color as your end product such as gray, white, or yellow.

Next, **heavy body acrylics** are applied in washes or dry-brushing methods to highlight the details carved into the foam (Figure **D**). It doesn't do any good to have a bunch of intricate details or battle damage if you don't make them pop.

You can have a great prop and a poor paint job can ruin it. Or you can have an OK prop and a fantastic paint job can save the day. While painting, I recommend having reference pictures available, to make sure you're getting the correct amount of contrast and weathering. With this armor, I took it a step further and added some **Fuller's earth dust** just to age the piece and give it another level of detail.

Check out this and the other free projects on my channel and start your journey to becoming a foamsmith! ●

PASSIVE MOTION MASKS

ADD MOVEMENT TO YOUR MASK — WITHOUT MOTORS

Written and photographed by Willow Creative

Masks are awesome and come in many shapes and forms, but adding special features will take them to the next level! Here I'll show you how I approach creating an advanced mask with moving parts, such as my werewolf head, with no motors required.

For any mask, I start with a 3D scan of my head, and lots of reference photos of the animal, and I think about my line of vision, how I'll see out of the mask — see-through eyes, or hidden mesh peepholes, or in this case, looking out the mouth.

Then adding moving features to a mask is, for me, a process of reverse-engineering:

• I sketch which part is going to move, in what direction, and where the start and end positions of the part are going to be.

• Then I think about how a rotation point or push/pull system could be used to move this part in the desired position.

• These points of rotation or push/pull systems will then be connected to my favorite method of controlling the movement: by moving my jaw (Figure **A**).

For my wolf mask's snarling lip, I ended up connecting a moving nose piece to the moving jaw, as it lined up well and works well with the expression of a wolf. The wolf mask eventually evolved with moving ears too, using a push-pull system with a thin steel wire cable to draw

WILLOW CREATIVE is a professional cosplayer from the Netherlands. For her, cosplay is the perfect outlet to combine creativity, fandom, learning new skills, and showing her craft to the world. willowcreativ.com

them back and an elastic band to pull them back into position (Figure **B**). The cable is an idea from bicycle brake cables, which are incredibly versatile — you can position and line up such systems at a lot of different angles, with different ways of control. This allowed me to rotate the ears differently to the movement of the jaw (Figure **C**).

By positioning the pivot points closer or further away, you can adjust how far the motion will move the final part. I had to 3D print and adjust the werewolf design a few times, changing the angles and mounting points of the various parts to make sure that it worked well. So the first mask just had a bunch of holes drilled everywhere to figure out the best angles. Sometimes I just use popsicle sticks and paint stir sticks with a few rivets to have an idea of where the movement will lead me!

All of these features will need to be integrated in the final look of your mask, so an idea on how to cover up your mechanisms is crucial to the final result. However, I think just the raw frame looks pretty awesome too!

For the final mask, which I sell as a DIY kit, I added motorized blinking eyelids. I ended up using CreatureCast (a neoprene-based rubber

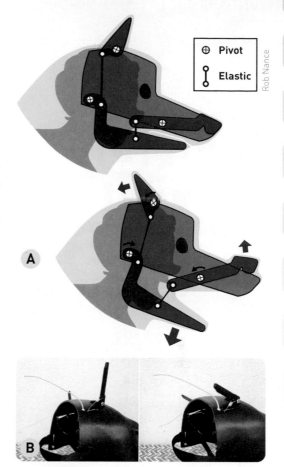

⊕	Pivot
⸙	Elastic

Rob Nance

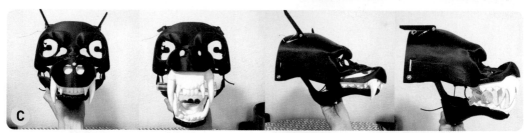

alternative to liquid latex) to create covers for the eyelid mechanism, and faux fur with plenty of allowance to create the rest of the skin without obstructing the movement. Remember to test your mask in both closed and open positions for all mechanisms, while applying the final covers!

For a more in-depth look at the function of the mask, check out the videos on my YouTube channel (youtube.com/@WillowCreative). Another interesting articulated mask is my Savathûn mask (Figure **D**), which utilizes a similar wire-pull system to create a very different look! ⊘

CANDLELIT CATHEDRAL DRESS

GO GOTH! SEW THIS SHOW-STOPPER WITH COLORFUL SHEER PANELS AND FLICKERING LEDS

Written by Christina Ernst

A

B

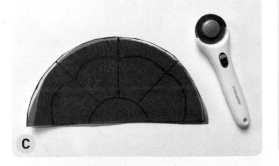

C

D

J.J. Xu, Christina Ernst

TIME REQUIRED: `2 Weekends`

DIFFICULTY: `Moderate`

COST: `$50–$100`

MATERIALS

» **Adafruit Gemma microcontroller**
» **Individually addressable WS2811 RGB LED bulbs, 1 meter string** such as Amazon B01M7MR44P, or eBay 374361502138
» **Lithium battery pack, 3.7V 2500mAh** with JST connector
» **USB data cable, Standard-A to Micro-B**
» **Double fold bias tape, black, ¼" wide (24yds)**
» **Translucent fabric, various colors (½yd each)** such as crepon or organza
» **Woven fabric, black (2yds) or premade skirt**
» **Lining fabric (2yds)** if not using premade skirt
» **Puffy paint or electrical tape**

TOOLS

» **Sewing machine**
» **Fabric scissors or rotary cutter**
» **Measuring tape**
» **Disappearing fabric marker (optional)**
» **Thick paper**
» **Sharpie**
» **Wire cutters**
» **Wire stripper**
» **Soldering iron**
» **Computer with Arduino IDE software** free from arduino.cc/downloads

CHRISTINA ERNST is a Chicago-based engineer, founder of shebuilds robots.org, and top prize winner of the Amazing Maker Awards 2022. make.co/amazing-maker-awards

I love adding special effects to textiles, and programmable LED strips have changed the game — but sometimes a look calls for more of a candlelit cathedral than a rainbow rave! This project uses flickering LEDs to illuminate semi-sheer fabric panels in the style of stained glass. You can upcycle an existing skirt, or sew your own garment from scratch.

Go for vibrant, jewel-toned fabrics — they pop the most when illuminated from behind. And choose bulb LED strips to flat ones; the light will project up into the whole window instead of out.

I love the intricacy of the center faux stained glass panel, and this project is infinitely customizable with your own panel designs. Mix and match, tell a visual story, or re-create your favorite glass art.

This project assumes proficiency with a sewing machine (straight and zig-zag stitches) and, if you'll make the whole skirt, with finishing a garment: specifically hemming, gathering, and adding a zipper or elastic waistline.

1. SKETCH STAINED GLASS PANELS

On posterboard or thick paper, draw a vertical line 5" shorter than the length from your waist to your desired hemline. This is your maximum stained glass panel height.

Sketch out your stained glass panels as desired and outline with Sharpie (Figure **A**). For the dress shown here, I sketched one central arched panel and four different, slightly smaller panels. You can optionally color the posterboard now to keep your pattern pieces organized later.

2. CUT OUT PATTERN PIECES

Use scissors to cut out every paper shape corresponding to a distinct color of fabric (Figure **B**). You'll also want a background piece the same size and shape as the entire panel (in this case, the entire arch).

3. CUT OUT COLORED FABRIC PANELS

Use fabric shears or a rotary cutter to cut a double thickness of organza or crepon for each piece (Figure **C**). The double thickness creates much more vibrant colors when illuminated by the LEDs and prevents the skirt from being too sheer. Repeat for every panel on the skirt.

4. SEW DOWN THE PANELS

Use a straight stitch in a neutral color to sew down each double thickness of colored fabric onto the background panel (Figure **D**). The raw edges will be covered when the black bias tape "lead lines" are sewn on top. Small visual gaps between pattern pieces are okay here since they'll be covered by the bias tape as well.

5. TRACE THE LEAD LINES

Piece together your original paper pattern pieces and place your sheer fabric panel on top. Use a fabric marker with disappearing ink or a very fine Sharpie to trace over the lead lines, pressing hard

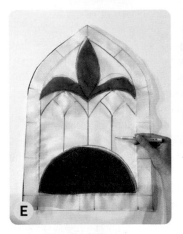

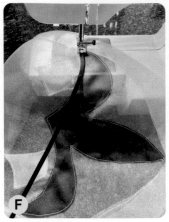

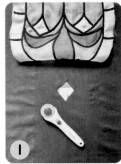

to transfer the ink to the fabric (Figure **E**). The black bias tape will be stitched over these lines to mimic the look of soldered panels.

6. SEW THE BIAS TAPE

Use a wide zigzag stitch with black thread to stitch the bias tape over all the lead lines you just marked (Figure **F**). The outer edge of the panel should not have any bias tape (Figure **G**). Snip any loose threads on the back of the panel so they don't show through the sheer fabric.

7. CUT THE SKIRT BODY (OPTIONAL)

If you're not using a premade skirt, calculate the size of your skirt panels based on your waist to hemline measurement L and your waistline measurement W. The skirt will be gathered from 3W to W for a nice full silhouette. Cut three panels from your black woven fabric (Figure **H**): one front panel that measures L × (3W/2), and two back panels that measure L × (3W/4). For the skirt shown here, the pieces were one 26"×44" panel and two 26"×22" panels.

8. ATTACH STAINED GLASS PANELS

Decide where on each skirt panel the stained glass panels should be positioned. Cut a hole about the size of a quarter in your black woven fabric in the center of where each panel will go (Figure **I**); this will make the next cutout step easier. Place each stained glass panel face down on the wrong side of your black woven fabric, centering the panel over the hole. Pin in place (Figure **J**). Use a short straight stitch to sew

Christina Ernst

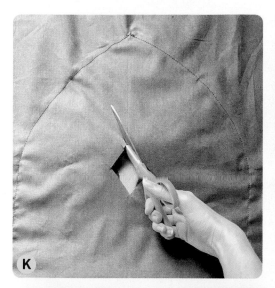

K

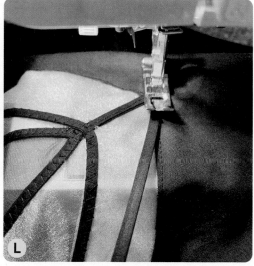

L

M

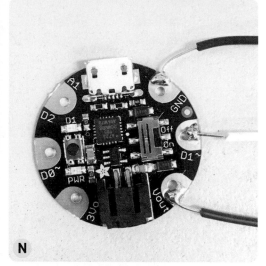

N

along the perimeter of the stained glass panel, including the bottom.

Flip the right side of your fabric face up and insert fabric scissors into the hole. Cut towards the edge of the stained glass piece, cutting as close to the straight stitch as possible (Figure **K**) to reveal the glass panel. Cover the raw edge of black woven fabric with zigzag stitched bias tape around the perimeter, excluding the bottom hem (Figure **L**).

Repeat for every stained glass panel. Finish the hemline with bias tape (Figure **M**). Gather the waistline using two straight stitches. The outer skirt is complete!

9. PREPARE THE ELECTRONICS

Solder power, ground, and data lines between the microcontroller and LEDs (Figure **N**). My LED wires have faintly colored print explaining which connection is which.

Trim the LED strip to the same length as the hemline of the skirt.

10. PREPARE THE LINING (OPTIONAL)

If you're not using a premade skirt, sew a skirt lining from the lining fabric with the same measurements you used in Step 7. Turn up the bottom hemline 1½", or roughly the same height as the LED bulbs. The goal is to make a pocket for

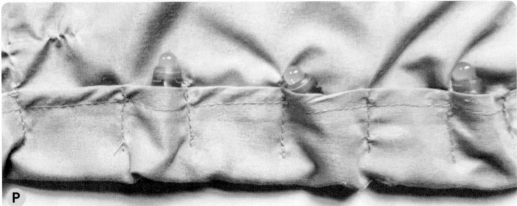

each LED bulb, securing it in place and holding it upright. Use pins to mark little pockets around each LED bulb (Figure **O**), then straight stitch along each pin for about 1". Leave the bottom ½" unsewn to accommodate the LED wire (Figure **P**). Hot glue also works as a convenient time saver.

Sew the lining to the outer skirt, with the LEDs in their pockets facing the backside of the stained glass panels. Finish the waistline by adding a zipper, elastic waist, or full bodice.

11. ADD A BATTERY POCKET

Cut a 10"×4" rectangle of lining fabric, and hem all four edges to create a rectangular pocket (Figure **Q**). Stitch the bottom and sides of the pocket down onto the skirt lining where the battery pack will be, leaving the top open for easy access. Place the pocket on the left or right side of the dress close to the hemline to best access the battery pack and to avoid sitting on it. Optionally add a snap or velcro to keep it secure.

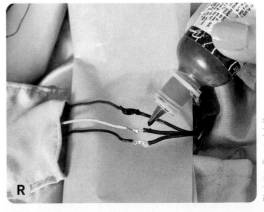

Christina Ernst, J.J. Xu

12. INSULATE WIRES

To prevent short circuits, insulate any exposed wires with puffy fabric paint (Figure R). Use wax paper or a drop cloth to protect your fabric from stains as the paint dries. Alternatively, you can use electrical tape to insulate the wires.

13. FLASH THE MICROCONTROLLER

Download the candlelight code file, *candlelight_pixels.ino*, from the GitHub repository at github.com/cornst122/candlelight_pixels. This code is a modified version of Neopixel Flames by Simon Cleveland (codebender.cc/sketch:271084). Using the Arduino IDE, open the file and select Adafruit Gemma 8MHz as the board, and USBtinyISP as the programmer. Gemma does not use a port.

Connect the Gemma board with the USB cable, then upload the code to the board.

> **NOTE:** Not all USB v3 ports recognize the Gemma bootloader. Use a USB hub or USB v2 port if the Gemma will not enter bootloader mode. If you see the error **A programmer is required to upload**, navigate to Sketch → Upload Using Programmer and use this option to upload instead of the IDE button.

The code is formatted for GRB pixels instead of RGB pixels. If you end up with green flames, switch the **g** and **r** variables in the code. Play with the following variables to affect the code in the following ways:

brightness (0-255): a higher value will be more visible, but will drain the battery faster.

r, g, b (0-255): these are the base color values that your flame will flicker away from.

color_delta: the degree to which a flickering flame strays from its base color. A bigger delta will have a more pronounced effect, but could stray into non-fire-toned colors.

Judge your candle flickering effect on the whole LED strip, not just one. What looks good on one LED looks chaotic across an entire strip.

14. FINISH AND ENJOY

Finalize the code, iron the dress, and insert a battery pack. Twirl and enjoy!

GLOW ALL NIGHT!

This dress looks and photographs best after sunset or in near darkness. Concerned about battery life? The rule of thumb for power draw is 20mA per pixel. This project's 2500mAh battery can theoretically support 30 pixels for about 4 hours. If you're carefully planning for an event, use the worst-case scenario (full white brightness) of 60mA per pixel.

Not interested in lugging around a battery pack and pixel strip all the time? Attach the LEDs, microcontroller, and battery pocket to a velcro-backed fabric strip. Use matching-sized velcro on the hemline of the skirt lining to easily attach or remove electronics. ⊘

WELDING LARGE 3D PRINTS

MAKE MASSIVE PROPS AND COSTUME PIECES BY MELTING PRINTED PARTS TOGETHER

Written and photographed by Dylan O'Connell

TIME REQUIRED: 10 Minutes

DIFFICULTY: Easy

COST: $20

MATERIALS
» **3D printed parts in PLA** to weld together
» **Spare PLA filament**

TOOLS
» **Soldering iron, wired**
» **Safety glasses**
» **Respirator**
» **Sandpaper, 80 grit**
» **Tape**
» **Cyanoacrylate (CA) glue and accelerator (optional)** aka super glue

As soon as you start your 3D printing journey you instantly feel like Tony Stark. Your mind starts to wander toward Iron Man suits and all the armor and giant props you'd love to have. Then you realize most printers are not able to print full body parts or 6-foot swords in one go.

So what do you do? *PLA welding,* a process where you use an ordinary soldering iron to fuse your printable-sized pieces together to create massive props!

1. SAFETY
Before you start, please put on your safety glasses and respirator. We will be melting PLA plastic and possibly super glue. The PLA, while

not inherently toxic, still should not be inhaled — and definitely not burning super glue!

2. SAND THE EDGES

Before you place two pieces together, make sure you have removed any support material or small imperfections on the edge to align them flush. Use your 80-grit sandpaper to knock down the edge (Figure **A**) till you have a great fit!

3. TACK-WELD PIECES TOGETHER

Before you go for the full-blown weld, align the pieces face down and touch the soldering iron to a small spot near each edge, lightly melting the edges of each piece together (Figure **B**). This will keep your pieces aligned while you continue the welding process.

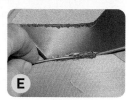

> **CAUTION:** Sometimes it's hard to hold pieces together and get them tacked. If you can't get by with taping them, instead you can use CA super glue to get the initial hold. However, in the next step you run the risk of melting the super glue crystal which gives off a terrible, toxic gas that will make you feel like you're going to go blind. So super glue is a last resort!

4. WELD

This is a very simple process. Your goal is to melt the edge of each piece along the seam, then mix the molten plastic to become one piece. Think of folding the edges of each piece on top of the other. Touch the soldering iron to the left of the seam, lightly melting the PLA, and then on the right side. With both sides at a pliable state use circular motions with the iron to mix the plastic from each side (Figure **C**). Continue down the seam till you're happy with the bond (Figure **D**).

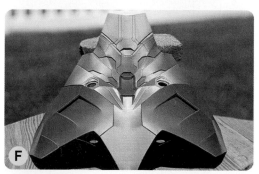

> **TIP:** If you have extra PLA on a spool you can use the extra filament to add strength to the weld. Place the strip of filament over your initial weld and then melt the filament into the seam (Figure **E**). This gives a nice sense of security to the weld.

With this technique you can create massive 3D printed parts. PLA welding creates a very sturdy structure by fusing the prop into one piece (Figure **F**). Soon you'll be carrying around a 6-foot Buster Sword (Figure **G**) like you've always dreamed of!

Beware, you will then be asked all the time, "How big is your printer?!" ⊘

DYLAN O'CONNELL (Quest4Nostalgia) is a YouTube content creator sharing tips, tricks, and tutorials for 3D printing movie props and replicas, from Mighty Morphin Power Rangers helmets to full Iron Man suits.

BLOOMING DRESS

THE ART OF ACTUATING LEAVES

Written by Rehana Al-Soltane

TIME REQUIRED: 3 Days
DIFFICULTY: Advanced
COST: $60–$80

MATERIALS

- » **High torque 6V servo** ROB-11965 from SparkFun
- » **Flexible filament, black** like NinjaFlex
- » **Satin fabric** in burgundy and green
- » **Monofilament fishing line**
- » **Acetate sheets**
- » **Needle and thread**
- » **Copper foil sheet** with conductive adhesive

FOR THE PCB:

- » **Copper plate**
- » **SAM D21E microcontroller**
- » **LEDs (3)**
- » **Resistors, 1kΩ (3)**
- » **3-pin headers**
- » **Capacitors, 1µF (2)**
- » **Voltage regulator, 3.3V**
- » **2×5 header**

TOOLS

- » **Sewing machine**
- » **3D printer** I used a Prusa i3 MKS3.
- » **PCB milling machine** Roland monoFab SRM-20
- » **Soldering iron and solder**

SOFTWARE

- » **Fusion 360**
- » **InkScape**
- » **SVG-PCB PCB editor**
- » **Arduino IDE**

REHANA AL-SOLTANE is an educator and creative technologist who works in the field of maker education and physical and computing. Her passion for tech education led her to Harvard and MIT, where she got her master's degree in Learning Design, Innovation, and Technology from the Harvard Graduate School of Education.

Alexia Asgari

p:
h/save50

BUSINESS REPLY MAIL
RST-CLASS MAIL   PERMIT NO. 187   LINCOLNSHIRE IL

POSTAGE WILL BE PAID BY ADDRESSEE

Make:
PO BOX 566
LINCOLNSHIRE IL 60069-9968

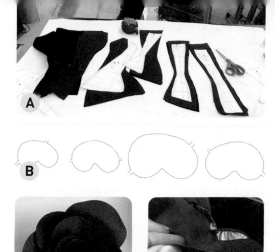

Since I was a child, I have been making (and breaking) a lot of stuff: at home I would sew miniature dresses for my dolls using needles, thread, and fabric; in my father's garage I would take apart car engines with a screwdriver and drill; and later I would build herds of robots out of potatoes and parsnips. A few months ago, I decided to enroll in the MIT Media Lab's "How To Make (Almost) Anything" crash course in digital fabrication tools, taught by Neil Gershenfeld. Combining those digital fabrication skills with the craftsmanship passed down to me through several generations of seamstresses, I created the Blooming Dress: a completely open-source wearable tech dress with actuating fabric roses that "bloom" in response to capacitive touch.

THE BODICE

I started out with the bodice using Aranea Black's free Dolores corset pattern (which at press time is no longer available), printing it on paper, taking measurements, and cutting my exact pattern (Figure A).

I hand pinned all the panels together first, then sewed them together on the sewing machine and added an open-ended zipper.

MAKING THE ROSES

For the roses, I used burgundy-colored satin fabric. I made a few test roses and through some experimentation with different shapes of petals, I decided on a heart-shaped petal in four different sizes: 2", 3", 4" and 5" (Figure B). I used Fusion360 to design the smallest leaf and then duplicated and expanded it. Because I was going to laser cut a whole bunch of these, I added little tabs to keep the fabric from flying off in the laser cutter.

To sew the actual roses, I used three petals from each size, starting with the smallest petal in the center, and then continued to wrap around the rest of the petals, while using a needle and thread to keep them together (Figure C).

I also made one smaller capacitive rose to act as a sensor, by cutting out a petal shape from adhesive copper tape that I stuck on the fabric petal (Figure D). Touching this capacitive rose would eventually trigger the actuation that makes the fabric roses bloom.

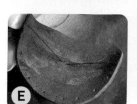

Rehana Al-Soltane

THE ART OF ACTUATING LEAVES

Actuation was unfamiliar territory for me, so I started searching for the type of actuation I wanted to achieve; it quickly became apparent that incorporating electromechanical structures into fabrics was not a well-explored area.

1. PROOF OF CONCEPT

I began experimenting with different techniques to mimic the actuating motion for the leaves by hand. From a foam sheet I cut a petal, and with a needle inserted thread through the top, in the middle at two close points, and through the bottom of the petal (Figure E). By pulling and releasing the thread I got a smooth and consistent motion of the petal opening and closing.

2. 3D PRINTING ON FABRIC PETAL

I set out to create another proof of concept that would work on fabric. I designed several 3D-printable horizontal channels that mimic the veins in organic leaves, and incorporated a hollow center to allow the thread through (Figure F).

I printed it on the test fabric using the 3D sandwiching method:

- Print the first 3 layers
- Pause the printer
- Lay down the fabric on top and clamp the fabric tightly to the bed using clips
- Resume printing

As a result of this technique, the lower and upper 3D printed layers merged together through the fabric, creating a strong and cohesive structure (Figure G).

At this point I also discovered fishing line as a much stronger alternative to regular thread, but upon testing it, it felt too flimsy: the leaf didn't return to its original state, mainly due to the friction between the line and the 3D print, but also due to the weight of the print on the fabric.

3. ACETATE SHEETS

I experimented some more. I sandwiched a foam piece between two satin petals, but this too prevented the fishing line from returning when released, due to too much friction. I ultimately decided to sandwich a thin sheet of acetate between two satin leaves that I cut in a heart-shaped form. I sewed these together using a sewing machine and with a soldering iron melted several holes diagonally on each half (Figure H).

I looped the fishing line through the upper hole first, and then through the middle and bottom holes. Each petal would have two separate lines that come together in the bottom center hole. By pulling the lines by hand, I got the exact actuation I had envisioned!

For the actual mechanical actuation, I used a 6V 360° servomotor, and threaded the fishing line through a hole in the servo horn that I attached to the servo (Figure I).

CREATING THE CHANNELS

Because I would have the servo hidden in a pocket on the hip of the corset, I needed something to guide all the lines from the flowers to the servo. I decided to design very thin, flexible channels with sewable tabs (Figure J), and make them as long as the 3D printer bed would allow (Figure K).

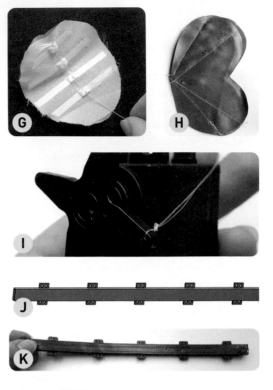

PCB DESIGN

Over the course of "How to Make (Almost) Anything," I made a ton of PCBs and fell in love with the entire process, from designing and milling to soldering. Part of why I found it so enjoyable is the software I chose: SVG-PCB, a PCB editor created by Leo McElroy and Quentin Bolsée that runs in the browser can import SVG files. You can edit your board design through both the code and the visual editor.

I wanted to design a one-of-a-kind PCB in the shape of a mannequin, so I created a silhouette in InkScape, exported it as an SVG file (Figure L), and imported it into SVG-PCB (Figure M). From there, I copied the SVG path and declared it in the **geo.pathD()** function. The final board uses a SAM D21E, two 3-pin servo headers and three LEDs. I also used vias to ground the ground pins of each servo header without having to jump over any traces.

I milled the board successfully on a Roland monoFab SRM-20 and soldered all the components on (Figure N). One thing I forgot to add to my PCB is a sewable pad for the capacitive rose, but I used a wire to jump to the rose instead.

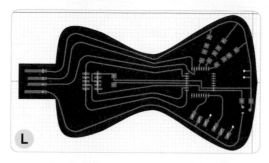

L

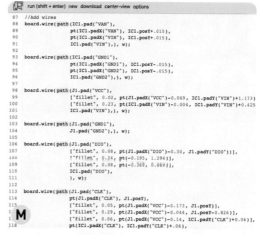

run (shift + enter) new download center-view options
87 //Add wires
88 board.wire(path (IC1.pad("VAN"),
89   pt(IC1.padX("VAN"), IC1.posY+.015),
90   pt(IC1.padX("VIN"), IC1.posY+.015),
91   IC1.pad("VIN"),), w);
92
93 board.wire(path (IC1.pad("GND1"),
94   pt(IC1.padX("GND1"), IC1.posY-.015),
95   pt(IC1.padX("GND2"), IC1.posY-.015),
96   IC1.pad("GND2"),), w);
97
98 board.wire(path (J1.pad("VCC"),
99   ["fillet", 0.02, pt(J1.padX("VCC")-0.069, IC1.padY("VIN")+1.173)],
100   ["fillet", 0.23, pt(IC1.padX("VIN")-0.004, IC1.padY("VIN")+0.425
101   IC1.pad("VIN"),), w);
102
103 board.wire(path (J1.pad("GND1"),
104   J1.pad("GND2"),), w);
105
106 board.wire(path (J1.pad("DIO"),
107   ["fillet", 0.08, pt(J1.padX("DIO")-0.30, J1.padY("DIO"))],
108   ["fillet", 0.24, pt(-0.195, 1.294)],
109   ["fillet", 0.08, pt(-0.369, 0.869)],
110   IC1.pad("DIO"),
111   ), w);
112
113 board.wire(path (J1.pad("CLK"),
114   pt(J1.padX("CLK"), J1.posY),
115   ["fillet", 0.05, pt(J1.padX("VCC")-0.173, J1.posY)],
116   ["fillet", 0.29, pt(J1.padX("VCC")-0.044, J1.posY-0.826)],
117   ["fillet", 0.06, pt(J1.padX("VCC")-0.14, IC1.padY("CLK")+0.06)],
118   pt(IC1.padX("CLK"), IC1.padY("CLK")+.06),

M

PUTTING EVERYTHING TOGETHER

Now that I had my bodice, roses, PCB, and leaves completed, it was time to put everything together. I decided that I would have two actuating roses on the neckline (Figure O). I sewed the flexible channel directly onto the bodice, right under the roses, and pulled the fishing line from the actuating leaves through the channel. Lastly, I made a little pouch from two strips of velvet for the servo right under the channel.

I then gathered together all the fishing line, looped it through one of the holes on the servo arm, and made a tight knot. I also attached the PCB to the bodice using a needle and thread. Finally, I attached a few more roses around the neckline and waist, and I sewed the capacitive rose on the lower back of the bodice (Figures P and Q).

THE CODE

I wanted the servo to be triggered by the capacitive rose, and with help from Quentin Bolsée I used the Adafruit FreeTouch library. Since the servo would be pulling the wires, I experimented with several values to turn the servo arm to certain degrees. In the end, I decided on turning the servo 180° followed by a pause, then reversing by 90° and pausing again.

I had a lot of fun bringing my love for electronics, textiles, and physical computing together in this project. I would like to spend more time refining the actuation of the flowers and incorporating moving electromechanical pieces into more textiles! ◙

Get the Arduino code and see the Blooming Dress in full flower at makezine.com/go/blooming-dress

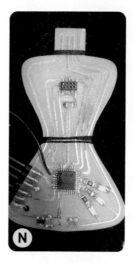

N

O

P

Q

Rehana Al-Soltane

NO-CODE LEDS AND ANIMATION

USE BEKONIX'S DRAG-AND-DROP TIMELINES FOR COSPLAY LIGHTS, MOTORS, AND AUDIO

Written and photographed by Ben Eadie

You've just 3D printed the most astounding prop, crafted an epic cosplay glow sword, or fashioned a head-turning trendsetter piece. But your masterpiece needs that little something extra to make it truly memorable. You're considering LEDs, but the thought of generic blinking lights feels underwhelming, and you don't exactly boast prowess in microcontroller programming ...

Fear not, fellow makers! This guide will introduce you to **Bekonix**, a no-code programming interface that's been a game-changer for me in the movie industry and made me a hero to my kids with cool LED effects for their projects.

DRAG-AND-DROP TIMELINES

Bekonix is a GUI-based *timeline programming interface* that's all drag and drop. Very rarely do you need to type any code out. No figuring out what coding function is needed, variables, enabling pins, none of that. You can program most things using only your mouse; the only time you hit the keyboard is to type in the name of the project to save it.

To help you grasp Bekonix's function, consider this analogy: Imagine a simple video editor. Now replace videos with LEDs, servos, buttons, and sensors. If you can trim down a basic video, you can program addressable LED lights with ease. Also, on the UI you can preview how your build will look and how it will react — the lights, buttons, servos, everything — without having assembled any electronics. So you don't even have to buy any components until you have the

BEN EADIE is a movie prop maker, practical special effects designer, inventor, nerd, and former aerospace engineer. You can see his work in films like *Star Trek Beyond*, *The Predator*, and *Ghostbusters: Afterlife*.

build all set up on your computer.

The cherry on top? You can connect and program your microcontroller over Bluetooth, no cables! Even use your smartphone or tablet as your remote control. Don't be deterred by the software subscriptions on the Bekonix site; their free plan is more than adequate for any hobbyist.

ADDING LEDS TO COSPLAYS

What can you do with this timeline interface? If you have an LED strip or matrix (Figure **A**), you can have it run an animation timeline (Figure **B**) over and over again — or have it turn on once and throb red to white with a switch, for example. Even better, once you get your teeth into this, you can set up inexpensive Grove sensors (Figure **C**) to make your LEDs react to noise in the environment, or light things up when a person approaches with a proximity sensor.

All of this is already in the programming environment. You just buy the recommended sensor, attach it to the pins laid out on the user interface, and hit play (Figure **D**). It really is that simple. Here's a video I did showing you how to make a *Ghostbusters* Ghost Trap light bar with the system: youtu.be/-ETc7D-sves

But there's more! In *asynchronous mode*, all your animations can run independently from each other, so if you want one part of a prop to light up but the other to stay dark until you hit a button, no problem. Or, and this is super cool, imagine you have an Iron Man helmet. You can run the servo to open and close the faceplate and have it run the glowing eyes and voice modulator too! Yes, you read that right, Bekonix can manage servos and audio functions, and they can all interact with each other as well. Once the faceplate is down, you can have it automagically light up the eyes. Really the only thing limiting you on this system is your imagination.

Whether you're comfortable with soldering and setting up RGB LEDs with a microcontroller, or you're a beginner who's not so keen on soldering, Bekonix has something for you.

• DON'T WANT TO SOLDER?

Start simple with the **nLiten** boards from MakeFashion (nliten.tech). Their Basic Dual Edge Kit is an all-inclusive starter pack

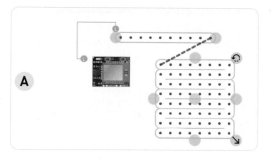
A

B

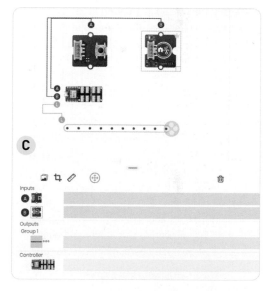
C

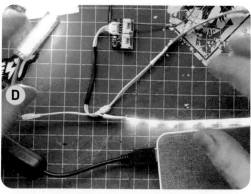
D

featuring LEDs, nLiten microcontroller, cables, and a USB power adapter. Just plug in the USB to a power source, connect the LEDs to the board, launch Bekonix, and you're ready to start programming! Bekonix and nLiten offer comprehensive online tutorials to help you navigate the system and kick-start your light animation journey. Check out the inspiring creations on the MakeFashion.ca site for a taste of the dazzling projects that have been crafted using this system.

Then check out Bekonix's Components Library (bekonix.com/platform/ componentsfeed) where they recommend plug-and-play parts — sensors, servos, motors, LEDs, switches, etc. — that you can use immediately. It's a great place to start.

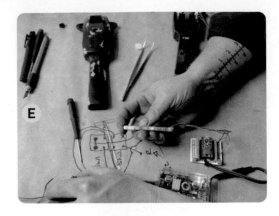

• DON'T MIND SOLDERING?

Consider investing in a **Seeed Xiao** microcontroller board and a strip of addressable LED lights. A little research will reveal that your total cost could be less than $20. Xiao boards are not only tiny and inexpensive, but they've also received rave reviews from *Make:* (makezine.com/ comparison/boards), making them a compelling option for the budget-conscious maker.

There's an extra step with Xiao where you'll need to upload the Bekonix OS backend system to the board via USB using the Bekonix app. It's straightforward and, after I've done countless boards, has proven to be reliable and super easy. Just boot the board, connect it to your PC, and open up Bekonix. Select the board in the drop-down to do an initialization process and it's done. Once uploaded, you can disconnect from the computer and program the Xiao via Bluetooth, just like in the non-soldering option. For all of this, you can get easy-to-follow tutorials at their website.

Within Bekonix, simply drag and drop the Xiao board into the GUI window as well as your LEDs. The system will even guide you on the correct pin connections for soldering. Then you're set — get creative and start layering various LED effects like crawl, pop, strobe, and more, using the user-friendly interface.

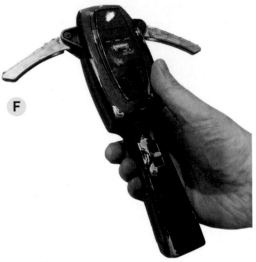

MY NEW GO-TO SYSTEM

What have I used this for? Ghost Traps (page 52), Proton Packs, and PKE Meters (Figures **E** and **F**)! I discovered this system after I worked on *Ghostbusters: Afterlife*; now any time I'm asked to make props with lights, it's my go-to system. Period. Here's a clip from a video I did for Ghost Corps after the release of *Afterlife* on how to take a Hasbro toy PKE meter and make it look 10x better using Bekonix: youtu.be/VbZ1Hjalevs.

As makers, we are in a constant quest for innovation and creativity. This quest often leads us to amazing discoveries like Bekonix that open up a whole new world of possibilities. With Bekonix, anyone can breathe life into their projects with captivating LED animations and sensors, motors, and servos. It's time for you to stop thinking "I can't do electronics" because now it is simply not true. You absolutely can. ●

COSPLAY PRO TIPS

STAR COSPLAYERS SHARE THEIR FAVORITE **TOOLS**, **TECHNIQUES**, AND **COMMUNITIES**

① Lady Jessica in Fremen stillsuit, from *Dune*.

① BEVERLY DOWNEN AKA DOWNEN CREATIVE STUDIOS

Portland, Oregon
downencreativestudios.com

• TIPS, TOOLS, TECHNIQUES:

I use **EVA foam** to create most of my props and costumes. My most important and favorite tool I use to transform this versatile material is my **Dremel rotary tool with a flex shaft attachment**. It makes the Dremel easy to hold while I'm manipulating and carving elaborate textures and details into EVA foam, and reduces hand and wrist fatigue when I'm working on a big project.

When trying a material or process for the first time, **experiment on scraps.** Keep a few scraps nearby and apply the same processes and products to those scraps as you work, being sure to document the steps along the way with photos and notes. These samples provide you with a place to safely experiment with products, processes, and techniques before applying them to your final project pieces. For instance, discovering a paint incompatibility is better on scraps than on your final project! You can then keep the samples for future reference, replications, paint touch-ups, and more!

• FAVORITE COMMUNITY:

Of course **SheProp,** a supportive and safe space for cosplayers and makers who identify as female, nonbinary, and the trans umbrella. This is my favorite space to connect with other creators from a diverse background of experience, abilities, and specialties. I founded this group several years ago and we now have over 5,000 members worldwide! Our community is very active online (Facebook and Discord) and our members host SheProp panels, meetups, and events at conventions around the world. No matter where you are on your maker journey, you are welcome to join our supportive group!

Downen Photography

2 TrisRex with his Xenomorph Empress Alien.

PROPS TO SHEPROP

SheProp is a safe space and supportive community for cosplayers and artists that identify with a marginalized gender, including women, two-spirit, and the trans umbrella. In the SheProp Community space, members can post work-in-progress images, ask questions, and fully engage with a robust maker community without fear of receiving unsolicited "advice" and degrading commentary, a common issue within most other online maker spaces. SheProp (sheprop.com) was founded in 2018 by Beverly of Downen Creative Studios.

2 TRISTIAN JOHNSON AKA TRISREX

Chicago, Illinois
linktr.ee/TrisRex

• TIPS, TOOLS, TECHNIQUES:

This year my favorite tools have been the **Dremel and soldering irons for detailing foam!** They make foam projects much faster and more versatile due to the endless types of textures they can create. I've also been using a lot of **garbage bags and shrink wrap** for creating wrinkles or more organic detailing. Melting them with a **heat gun** can create some cool effects!

Create your own characters. The Empress Alien is my take on an older, more evolved version of the Queen Xenomorph from *Aliens* (1986), the movie that started this journey for me as a child. She's my favorite creature, beautiful and majestic but terrifying at the same time. When I started my *Aliens: Resurgence (For The Hive)* universe I knew I had to tell a new story of the Queen — to pay respect to the past, acknowledge the present, but give audiences a look into the future of how an evolved Queen can truly look practically.

3 Cal Kestis from *Star Wars: Jedi Survivor*.

4 Regency-inspired Cruella DeVille.

• **FAVORITE COMMUNITY:**
The **Stan Winston School of Character Arts** is a really cool resource — now we can learn how our favorite creatures were made, from the artist who made them. I try to apply the course I'm taking to whatever projects I'm working on, which has been very helpful for me. It's also been cool to connect with other monster makers and FX artists because at times it seems like an art of the past ... but we're still here!

3 JOSHUA DUART AKA ARTEMISTYCK

Atlanta, Georgia
linktr.ee/artemistyck

• **TIPS, TOOLS, TECHNIQUES:**
When sewing a new costume, always always *always* **make a mock-up** in a fabric similar to the final fabric you'll be using. It will allow you to fix any fit issues and make sure proportions are correct, etc.

• **FAVORITE COMMUNITY:**
Dragon Con! I have met so many incredible makers, professionals, industry folks — plus made a lot of my closest friends at this convergence of geeks, cosplayers, and enthusiasts. It's a great environment for creativity and fun to celebrate your favorite fandoms!

4 TIPHANI DIXON AKA CON, PROS AND CONS COSPLAY

Kansas City, Missouri
meetprosandcons.com

• **TIPS, TOOLS, TECHNIQUES:**
Really get to **know your own learning style** and approach cosplay crafting in a way that is in alignment to your way of processing information. There isn't a right or wrong way. If an application-based, "get your hands dirty" approach is the way you learn best, and that's what motivates you, start there! Or if you need to do research and understand the ins and outs of a project before attempting it, take your time and learn the theory before jumping in.

Learning to **style your wigs** will take your cosplay to another level! It can be difficult, but it really brings your costume to life. For styling lace front wigs, I love the **latch hook style ventilating tool** — much easier to control than the needle-style ones. It's very similar to making latch-hook rugs (something I did a lot as a kid).

• **FAVORITE COMMUNITY:**
I love spaces like **POC Cosplayers** on Facebook and **SheProp** that advocate and celebrate marginalized cosplayers and empower us to bring our full identities into the community. And many of my friendships have been forged at **cosplay contests,** backstage in the "green room" waiting to hear

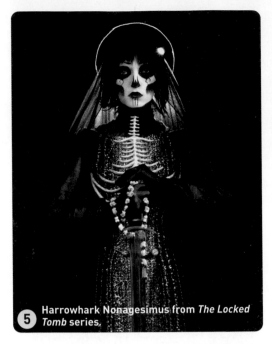

5 Harrowhark Nonagesimus from *The Locked Tomb* series.

6 Black Knight from *Monty Python and the Holy Grail*, with party streamers for gore.

results. You're there to compete, but it is also a great time to meet, visit, and learn from other makers, and celebrate the art of creating together. It takes a certain kind of person to pour that much time, energy, and effort into a costume, and it's wonderful to share that feeling with others.

5 REGAN AND SCONE AKA COWBUTT CRUNCHIES

Boston, Massachusetts
cowbuttcrunchiescosplay.com

• **TIPS, TOOLS, TECHNIQUES:**
Most cosplay wigs are made of plastic, so your number one wig styling tool should be **heat, not hairspray!** Heat styling lasts forever, unlike temporary glues, which can fade after a few hours. So add a little heat from a **blow dryer, steamer, or flat iron** to soften those wig fibers and shape them into a new, permanent direction!

• **FAVORITE COMMUNITY:**
We love original cosplay design, so we started our own crafting celebration: **The Cosplay Couture Gala!** This event features a design prompt, a makeshift runway, and our favorite part: the social mixer. It's a great opportunity to highlight talented designers in the cosplay community, admire their incredible outfits, and make new friends!

6 STEPHANIE CHAN OF FOAM ARMORY

Calgary, Alberta, Canada
foamarmory.com

• **TIPS, TOOLS, TECHNIQUES:**
I carry around a **travel size bottle of contact cement** with some cheap **sponge brushes**. It's surprising how often I pull it out for all kinds of repair assistance and sudden cosplaying on the fly!

Last year I wore my Black Knight costume as normal, but this year on a spontaneous lark, I instead used **party streamers to simulate the gory blood pouring effect!** I hopped around a recent con on one leg and the response was exactly what I was hoping for: laughs! Our company, Foam Armory, makes the lightweight **EVA foam chainmaille** I used for the maille coif.

FAVORITE COMMUNITY:
Evil Ted's Foam Fanatics on Facebook.

7 PHILIP ODANGO AKA CANVAS COSPLAY

Louisville, Kentucky
canvascosplay.com

• **TIPS, TOOLS, TECHNIQUES:**
Don't be afraid to flat pattern. If you're not comfortable draping fabric to visualize a design,

you can take measurements and plot a pattern on paper. **Gift wrapping paper** is helpful since the reverse side often has 1" graph line marks.

• FAVORITE COMMUNITY:

I started the Instagram online community called **Bros Who Sew** to empower and amplify male and male-identifying sewists. Posts shared with the hashtag #BrosWhoSew often include sewing tutorials, patterns, and finished projects.

8 JEN SCHACHTER AKA SCHACATTACK

San Francisco, California
jenschachter.com

• TIPS, TOOLS, TECHNIQUES:

Play with scale and contrast in your designs — don't be afraid to **push things a little farther than you think you need to** for impact. This is a trick I learned from watching Adam Savage work. Post production can certainly work magic for color and contrast, but things tend to seem flatter and less punchy on camera, or from a distance (like on stage). My take on Marie Antoinette is almost a caricature — sort of an 80s glam/Versailles mashup with a massive, larger-than-life wig and poufy sleeves.

Once you get the broad strokes of your design, you can start going in and adding little details, textures, highlights, embellishments that viewers discover as they come closer. But **be judicious about details** — everything at a high level of detail means no part really pops.

• FAVORITE COMMUNITY:

Halloween costume contests! An obvious choice, but this was certainly the gateway for many pro cosplayers. I have seen some of the most outrageous and clever ensembles made by novice crafters out of household materials. The ingenuity and playfulness always inspires me. ◓

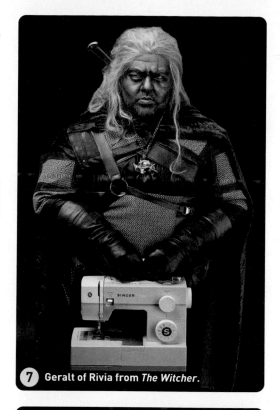

7 Geralt of Rivia from *The Witcher*.

8 Marie Antoinette EVA foam wig.

Love costumes and props? Get more tips, tutorials, and projects in our free Cosplay Collection PDF: make.co/cosplayprojectpremium

Alexandra Lee Studios, Stephanie Chan, David Morel/Singer Sewing Company, Norm Chan

Axiom Space

MAKING NASA'S NEW MOON SUITS

HOW AXIOM COMBINES SEWING AND 3D PRINTING TO CREATE THE NEW ARTEMIS SPACESUITS Written by Erin Winick Anthony

The eras of human spaceflight are visually defined by spacesuits. The bright white Apollo suit on the lunar surface, the silver of the Mercury pressure suits, and even the bright orange "pumpkin suits" of the Space Shuttle era created definitive looks for space exploration.

As we enter NASA's Artemis era, a new spacesuit will define humanity's return to the Moon. That suit is being created by Axiom Space in Houston, Texas.

The Artemis III mission (nasa.gov/feature/artemis-iii) will land the first humans on the lunar surface since Apollo 17 in 1972, so Axiom has more than 50 years of innovations and technological advancement to draw from.

"These suits are very much advanced over what was created in Apollo, but that is no slight to the Apollo engineers," Axiom's EVA deputy program manager Russell Ralston said. "Knowing what it takes to build spacesuits, I'm amazed still at what they were able to accomplish."

Axiom unveiled their first Artemis prototype suit to the public in March 2023. The goals of the new design were to provide increased flexibility, greater protection against the lunar environment, and more scientific tool options. Axiom says the new suit will allow for:

- an increased tolerance to lunar dust
- astronauts to walk longer distances and better manipulate objects thanks to a combination of soft and hard joints
- astronauts to walk on the cold surfaces of the Moon at its south pole
- more size and adjustability options to fit a wide range of crew members
- increased safety with multiple levels of redundancies
- HD video to be captured from the suit helmet

The design team took the lessons learned from legacy spacesuits and created a 21st-century design using new technologies like computer aided design (CAD) software and 3D printing combined with traditional techniques like sewing.

"We start with the design in the digital world, and depending on what the part is, we'll produce that either through 3D printing or soft wear manufacturing and patterning and stitching together," Ralston said. "Then we'll go test it."

The glove's exterior surface has multiple materials for optimal dexterity and grip.

The outer layer acts as containment and a heat shield for the inner layers. It looks pretty good too.

This complex structure allows for the elbow to flex in a wide range of motion.

ERIN WINICK ANTHONY is a science communicator living in Houston, Texas, and the founder of STEAM Power Media. She has a degree in mechanical engineering and formerly worked for MIT and NASA's International Space Station program.

Material sciences play as much of a role in spacesuit design as construction techniques.

Prototyping with a standard FDM printer during the early stages of the project.

An extensive list of materials are utilized to reach all the goals of the suit.

For spacesuit manufacturing, the tolerances required in sewing are extremely tight.

Ralston credits tools like 3D printing with allowing them to create this suit comparatively fast. It was crucial for quickly iterating on the designs. He estimates from starting the design process to a prototype suit sitting in the lab took his team only 8–9 months, compared to years for past suits.

"Portions of the backpack that you see on that suit were 3D printed. And just to be clear, that's not just because it was a prototype," Ralston said. "We do have 3D-printed, additively manufactured parts in our flight design. We will have some printers in house that we'll be using to produce those parts certified for flight."

The final design will have both 3D-printed parts that are internal to the space suit, and some that are exposed to the space environment. Based on the needs of each part, Axiom uses everything from inexpensive Prusa printers to much more expensive printers that can print reinforced plastics and metals.

"The print parts usually take a day or two, but some can take longer if they are particularly large or complex," Ralston said. "Most of the printed parts have loose tolerances, but if we need tighter tolerances we machine them after the print process in some cases."

All parts printed for spaceflight must follow NASA's technical standard for additive manufactured parts (NASA-STD-6030) which outlines all requirements for using 3D printing in the design, fabrication, and testing of parts for crewed and uncrewed missions.

So is this really a black spacesuit? No. The prototype's black cover is shielding some of the proprietary aspects of the suit underneath. The final product will be a necessary white color to assist with temperature control.

However, the creation of the cover was an opportunity for a sci-fi collaboration. Costume designer Esther Marquis from the Apple TV+ series For All Mankind worked with Axiom to develop the sleek black look.

"One of the things I love about working on spacesuits is it is this perfect blend of art and science," Ralston said. "While there is a lot of

This team comes from all backgrounds to collaborate on a tightly integrated, robust solution.

The suit is a feat of ingenuity and skills from many industries: military, aerospace, ballet, and theater.

science behind designing a spacesuit, especially for a lot of the life support equipment, when we make what we call the pressure garment, there's no textbook. There's no scientific formula that's going to tell you exactly how to design that to fit all these different people."

Spacesuits are a cross between a personal spacecraft and clothing, making sewing an essential piece of the manufacturing puzzle as well. But the sewing skills required for this type of creation are unique.

"Spacesuit stitching and manufacturing is more precise than anything else out there," Ralston said. "We've yet to find another industry that holds the tolerances that we need and has the same level of quality that we require for making spacesuits."

This carefulness is evident when you walk into their sewing labs. The labs are filled with single needle, double needle, off-arm, post, bar-tack, serger, and zig-zag sewing machines, all used for the creation of the suits. In typical clothing factories, the buzz of machines is constant and

fast. Axiom's sewing lab is almost dead silent. Some of the sewers even turn the machines by hand to achieve the level of precision needed.

These expert sewists are recruited from many different backgrounds to bring their skills to the space industry.

"We've got people that worked in the ballet and theater industry. We've got people that worked on space suits previously. We have people that worked critical military sewing applications, such as parachutes or things like that. They come from all walks of life," Ralston said.

Zach Paugh is one of the Axiom sewing techs who worked on the suit, and cannot wait to see his creation launch.

"It's like a little bit of each of us is going up there with the astronauts, and a little bit of our mentors, a little bit of our family," Paugh said. "It's more than just ourselves. It's everyone before us and everyone after us." ⊘

Axiom Space

TIME REQUIRED: 15–30 Minutes

DIFFICULTY: Easy

COST: Free

MATERIALS

» **Large cardboard box** Recycled is best, and the bigger the better.
» **Empty soda pop can**
» **Thumbtack or small brad nail**
» **Opaque tape** 2" black duct tape works great.
» **Sheet of white paper (optional)**

TOOLS

» **Box knife or scissors**

AUDREY LOVE is a creative technologist in Los Angeles, California, focused on storytelling and design. echoechostudio.com

CARDBOARD BOX
ECLIPSE VIEWER

MAKE A PORTABLE PINHOLE PROJECTOR FROM HOUSEHOLD STUFF TO SAFELY WATCH SOLAR ECLIPSES Written and photographed by Audrey Love

Two solar eclipses are coming to America: October 14, 2023, and April 8, 2024. When a total solar eclipse happens, the moon completely blocks the sun, making the sun's corona briefly visible. This is a rare cosmic treat to observe! But you can't safely look directly at the sun, even with sunglasses — you need an eclipse viewer.

Here's one that's easy to make from stuff you have around the house: a box pinhole projector. It's like a pinhole camera, except your eyes are the film and your brain is your photobook.

1. FIND A BOX

Look for a biggish box. The longer your projection is, the larger the projected image of the sun will be. If you use a box as long as your forearm or longer on four sides, you'll have great results!

2. CUT HOLES IN THE BOX

Cut a small rectangular hole in the box, just above center on the short side of the box (Figure **A**). This is where you will attach the pop can "lens" of the viewer.

On the box edge below, add a hole for your eyes. The hole should be slightly smaller than the width of your head, but if you cut it too large (like we did initially) you can add some tape back to the sides of your hole to darken the viewer and fit it to the sides of your face.

3. MAKE THE PINHOLE "LENS"

Using your box knife, cut from the pop can a square of aluminum that's slightly larger than the square hole you previously cut in the viewer box (Figure **B**).

You can also try to make a lens with layers of aluminum foil and scissors if you like, but I find that this lens is too delicate and flimsy when transporting the viewer. If you do use foil, try and get it as taut and flat as you can.

After the lens is cut, tape the aluminum over the rectangular hole in the box with opaque tape.

Using a thumbtack or tiny nail, poke a tiny hole in the aluminum (Figure **C**). This hole should be very small, not even the full diameter of the tack or nail used to puncture the aluminum. Hooray! You just made a rudimentary camera obscura!

4. FIX LIGHT LEAKS

When you have the viewer pulled up to your eyes, you won't want any light leaks ruining your view. I was able to block all the light leaks with black duct tape, but you can also use electrical tape or gaffer's tape. Any completely opaque tape, or enough layers of semi-opaque tape, can help solve your light leaks (Figure **D**).

5. CREATE A VIEWING PLANE (OPTIONAL)

By taping a sheet of white paper on the inside of the projection side of the box you can create a clearer viewing plane than just the regular brown cardboard.

Think of how films are projected at the movie theater. Theaters aren't projecting on a soft, dark surface, they use a taut, plain white background. A sheet of paper can also make it easier to take photos of the projection with a digital camera from inside the box.

GO SUN SEEKING!

Raise the box viewer to your eyes and stand *with your back to the sun*. Position the viewer until the bright dot of the sun comes into view. At eclipse time, you'll see the moon's shadow slowly block out the sun (Figure **E**). Enjoy making DIY science memories in the sun! ⊘

• • • *Turn the page for more eclipse projects* • • •

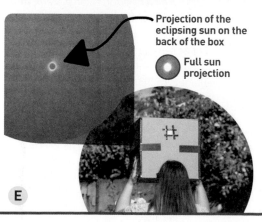

Projection of the eclipsing sun on the back of the box

Full sun projection

FOR ALL HUMANKIND

SPACE IS THE PLACE — FOR THESE COOL PROJECTS!

Will you serve on the bridge or lower decks? The fab, the pad, the orbiter, or the hab? Humanity needs dreamers and makers to chart our next chapter in space. Here are 20 fun projects you can make — and one party invite!

1 YOU ARE IN MY SYSTEM

Netherlands maker Illusionmanager created this wondrous orrery, a palm-sized solar system, from laser-cut wood, toothpicks, and a tiny stepper motor. With an ESP32 as its brains, the **Grand Planet Spinner** retrieves NASA data daily and has a nifty web interface — toss in any date and it will move the planets to their respective positions. instructables.com/Grand-Planet-Spinner

2 JOE STICKS THE LANDING!

In June 2019, Joe Barnard reckoned he was just months away from vertically landing his **Scout thrust-vectored model rocket** like Elon does. "I refuse not to stick the landing in 2019," he told

Make: (Volume 69, "Fly Like SpaceX"). Three years later, he achieved it (youtu.be/SH3lR2GLgT0). Congratulations Joe! Thanks for sharing your saga and your designs so others can try too. bps.space

3 PRINT A SATURN V

Relive Apollo greatness with a **3D-printed Saturn V rocket model** (makezine.com/projects/recreate-the-saturn-v-rocket-in-cad) from CAD Class (cadclass.org). Design it yourself in Fusion 360 with their free preview class, or just grab the 3D file and print one. There's even a version you can launch with our **Compressed Air Rocket Launcher**! makershed.com/search?q=rocket

4 OVERVIEW EFFECT

Like your own starship porthole, Matt Gray's **EPIC Satellite Photo Frame** streams each day's images of Earth from NASA's Deep Space Climate Observatory to a cute Pimoroni round display. And they're always on the sunny side because

DSCOVR orbits the L1 Lagrange point! Follow Matt's build and BOM at youtu.be/q4M3cm4ml_c and get the code at github.com/MattGrayYes/epic.

5 SPACE PARTY!

Reinvented is our fave new magazine, celebrating women in STEM fields. Their annual **Space Gala** is November 11, 2023 at Kennedy Space Center with music, maker activities, dinner, and dancing in the Rocket Garden. Guests include amazing maker Xyla Foxlin, astronaut Katya Echazarreta, Blue Origin and NASA engineers. Why not catch **Maker Faire Orlando** Nov. 4–5 too, and maybe a **rocket launch!** reinventedmagazine.com/space-gala

6 BACK-TO-BACK SPECTACULARS

America will witness the **next *two* solar eclipses**, an annular "ring of fire" October 14, 2023, and a total eclipse April 8, 2024. Get ready at eclipse. aas.org and solarsystem.nasa.gov/eclipses.

Annular Eclipse:
October 14, 2023

Total Eclipse:
April 8, 2024

PINHOLE VIEWERS — easiest and safest
Basic pinhole box: make ours on page 64!
NASA 3D-printed and laser-cut pinhole viewers: solarsystem.nasa.gov/resources/2921
Whole-room camera obscura: instructables. com/Giant-Camera-Obscura-Sun-Observer

SOLAR PROJECTORS — lenses and/or mirrors
Cheap magnifier: richardsont.people.cofc.edu/ safe_solar_folder and instructables.com/Solar-Eclipse-Viewer-From-Reading-Glasses-and-Card
Add mirror for bigger projection: instructables. com/Hardware-Store-Solar-Eclipse-Viewer
With camera lens: instructables.com/Sun-Spotter
With binoculars: instructables.com/Sun-Viewer-Observe-the-Sun-safely-with-binoculars

DIRECT VIEWERS — special solar filter materials
With solar filter film: from trusted vendors at eclipse.aas.org/resources/solar-filters
With welding glass: instructables.com/Simple-Affordable-High-Quality-Solar-Eclipse-Viewe
Giant eclipse glasses: Adler Planetarium made super-size specs in Chicago in 2017, and now Perryville, Missouri has made a pair for 2023–24. Scale up a real pair, cut from plywood or metal, paint, and add lots of solar filter film. Dimension drawings at makezine.com/go/giant-glasses.

SONIC "VIEWER" — for the sight impaired
LightSound: solar eclipse sonification tool converts light intensity into audio tones. Hear it and make it at astrolab.fas.harvard.edu/ LightSound.html

SOLAR FILTERS — for cameras and scopes
For camera: instructables.com/Cheap-and-effective-filters-solar
Telescope: instructables.com/How-to-make-a-solar-filter-for-a-telescope
Binocs: instructables.com/Binocular-Solar-Filters
Smartphones: Solar Snap lens and app, modify two phones for $13, eclipseglasses.com

PHOTOGRAPHING ECLIPSES — pro tips
SLR/DSLR/mirrorless: space.com/how-to-photograph-a-solar-eclipse
Plus digiscoping: bhphotovideo.com/explora/ photography/tips-and-solutions/how-to-photograph-a-solar-eclipse ⊘

JEREMIE BOULIANNE is a longtime maker professional, working for DIY companies such as Solarbotics and EZ-Robot in Calgary, Alberta. A *Make:* reader since Volume 02, he has a passion for electronics, 3D printing, and robotics, and he loves skateboarding and geocaching in his nonexistent free time. linktr.ee/botmatrix

GADGET CACHE ROAD TRIP!

Solving the clever contraptions of Trycacheus, in the "Disneyland of geocaching" **Written and photographed by Jeremie Boulianne**

Last summer I took an epic geocaching road trip to North Dakota. I drove past the majestic painted hills in Theodore Roosevelt National Park, across the badlands, past the colorful fields, all the way to the east side of the state to a one-horse town called Gilby. Gilby may be small but it's mighty! It contains a rare treasure for those discerning enough to look past the farmers' crops and into the bushy windbreaks between them. It has been hailed as "the Disneyland of geocaching" and it definitely lives up to the name.

First introduced to me by Joshua Johnson, aka The Geocaching Vlogger, Gilby is considered a geocaching mecca because of an innovative and creative geocache builder named Chad Thorvilson, better known as Trycacheus. He installs caches throughout his community wherever landowners will agree to host them. For those not already well-versed in the art of outdoor treasure hunting, **geocaching** is a real-world outdoor game of adventure and discovery. Participants use the Geocaching app or a hand-held GPS to find and log hidden caches of varying size and complexity all around the world. It's highly likely that there are some geocaches hidden around your present location.

Trycacheus specifically builds a rare and special type of geocache called a **gadget cache**. Gadget caches are the Holy Grail of geocaching. Dense collections of them, like in Gilby, are few and far between. The only other places I'm aware of that have such a concentration of them are Martinsburg, West Virginia, and Seattle, Washington (see page 72). I also hear there are quite a few in Germany.

Gadget caches are a high-tech twist on the traditional geocache-finding experience; cachers must also solve a challenging physical puzzle before they can access the logbook and record their find. Gadget caches are like a locked-room puzzle, but hidden outdoors, built tough and rugged enough to withstand the elements. They are a significant challenge to dream up and build as they must be designed to take abuse from both nature and cachers alike.

Trycacheus' gadget caches are remarkable. No two alike, they are some of the most well-crafted, high-quality, and heavily favorited geocaches I've ever seen. They include electro-

mechanical puzzles, mind-bending mazes, and challenges that demand physical dexterity and patience. From swinging large wooden mallets, to rearranging magic blocks, to controlling a robot arm (my personal favorite), each gadget cache is an intriguing and unique experience. After 2 days of walking narrow paths, ducking under twisted trees, dodging mud bogs, and stepping over thickets, I found and solved 30 of his gadget caches and I still didn't catch 'em all.

The highlight of my trip was spending a few hours with Trycacheus. He is a kind and genuine maker with a background in carpentry. I was honored to get a behind-the-scenes tour of his process and mind-blowing shop. Imagine cache building materials piled to the rafters, as well as shelves, drawers, and bins full of hardware, switches, and electronic components. Add to that a full woodworking setup and copious tools, and it's a maker's paradise.

So the next time you're craving an adventure, grab a GPS-enabled device and head to Gilby for the gadget caching experience of a lifetime. Unlike the real Magic Kingdom, there's no entry fee, and you can skip the line whenever you want. You'll be challenged, impressed, and inspired, all while battling the unknown dangers in the "wilds" of North Dakota. I only hope you'll enjoy it as much as I did.

① REALLY FUN INTERESTING DEVICE?

The first gadget cache I encountered in Gilby had a lot of red herrings. Taking the first letter of each word in the title, you can figure out what technology might be inside to unlock the goodies!

② LARRY HAS RETURNED

This type of puzzle is called a *cryptex*. You have to line up the letters in just the right position to open it, like that bike lock we had as kids — except this one has an interesting trick to it.

③ STRONG MAN AT THE FAIR

One of Trycacheus' most famous caches, with 350+ favorite points. It's a ton of fun and works just like the game you'd find at the fair. It's a test of raw strength and precision, and only the mightiest will be rewarded with access to the cache!

④ YOU BE THE JUDGE

Instantly my #1 favorite cache of all time — you get to control a robot arm to try and find a micro sized cache secretly hidden under one of the random objects inside. Each switch controls one of the robot motors and you need patience and precision to be successful.

⑤ BULLDOZER

Surprisingly fun, and has some twists to it. It uses magnetism in interesting ways. I want to make a version of my own!

⑥ E.T. RIDES AGAIN

The object is to generate voltage with a dynamo attached to the exercise bike to power a solenoid to release the trap door to get at the cache. But — the solenoid will only activate with the proper knife switch positions. Word to the wise: Don't touch the switch contacts while pedaling, you might get a tingly feeling in your hand like I did!

⑦ I SEE BM

This unique cache had a few tricks to it. I've never encountered anything quite like it. This cache will pump you up!

⑧ MAGIC BLOCKS

A clever electronic puzzle that requires you to set the magic blocks side by side in the proper order to get the combination for the padlock to show up on the 7-segment displays. Build one on page 74!

⑨ LAST STOP SALOON

Requires you to count a number of specific items in order to find the digits to enter in the proper code on the number pad. It usually takes a few tries to get the counts right.

⑩ THAR SHE BLOWS

Dispense yourself a clean straw and then insert it into the small holes and blow! You need just the right amount of air pressure to make the lock digits appear.

⑪ A CACHE YOU CAN'T RESIST

This gadget is one that many makers will appreciate, especially the electronics types like me! You get to learn about the resistor color code and reading values. Highly educational and fun.

⑫ PULL MY FINGER

Makes a familiar sound each time you pull a finger, haha! There's a super secret trick to solving this cache, which did take me a bit of time — one of those caches where you have a single "aha!" moment.

⑬ FTF CACHE

Cachers love to be the "first to find" a cache right after it's published — a coveted achievement in the geocaching world. This cache gives out FTFs like they're candy.

⑭ MICRO AMMO CANS

I kept stumbling across these cute 3D-printed micro-sized ammo cans. I traded some trinkets for one, because I had to have one. Later I was lucky enough to run into the maker of this tradeable treasure, Lisa Kaufman (GmaHat); she found the ammo cans on Etsy, stickered them, and loaded them with miniature swag. We shared some great Gilby gadget caching experiences! Funny enough, we were even coincidentally staying at the same hotel that weekend. ◐

GEOCACHES ARE ALMOST EVERYWHERE

There are over 3 million geocaches hidden in almost every country on Earth. There's even one geocache in space, on the International Space Station. —*Geocaching HQ, Seattle*

Super Hero's in Jeopardy

Coffee Break

GADGET CITY

Meet BounceBounce — undisputed champ
of urban gadget caching   **Written by Chad Champion**

Salmon Run

It Takes Three

Chad Champion

When I think about geocaching in the Pacific Northwest, images come to mind of hikes through dense green trails, ammo cans disguised by natural moss, and bison tubes strategically placed on branches. So as a hider, I've tried to brings my hides out of the woods and into an urban setting, just miles from Geocaching Headquarters (geocaching.com) in Seattle, Washington.

My goal is to elevate the concept of geocaching through urban hides. These tend to be so well camouflaged that I add notes in the Geocaching app description assuring cachers that what they've found at ground zero is indeed a geocache! Often I confirm the searcher's location with signs or

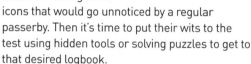

icons that would go unnoticed by a regular passerby. Then it's time to put their wits to the test using hidden tools or solving puzzles to get to that desired logbook.

Joshua The Geocaching Vlogger has called me the Greatest of All Time (GOAT), a title I try to live up to with each hide. Most of my caches are two-step — that is, once you solve the puzzle and open the cache, you have another puzzle to solve in order to obtain the logbook and earn that smiley.

Even more than my hides, I love inspiring others to build their own gadget caches. I've partnered with cachers throughout the USA and Canada (and soon Germany), and I host a podcast, *Gadget Talk*, with fellow gadget builder Baker6Clan (Derek Baker) where we teach how to build gadget and creative caches.

Want more hands-on learning? Find me at Gadget Academy (gadgetacademybuilds.com) at Geocaching Mega-Events each year where you get an opportunity to develop skills like soldering and walk away with a new puzzle to use with a cache of your own. We liked Trycacheus' Magic Blocks cache (see page 71) so much that we made it into a kit. You can build it on page 74.

Here are some of my urban gadget caches you can find in the Seattle area, and some hints on how they work.

① SUPER HERO'S IN JEOPARDY

What appears to be a pump station in a loading dock yields a trivial surprise. Using a keypad to solve the first question to open the pump station, you find yourself in the world of *Jeopardy* — you must answer randomized questions in three categories with RFID cards. Upon successful completion of Final Jeopardy, a hatch will open containing the logbook.

② COFFEE BREAK

Who doesn't appreciate a well-guarded cache? This security camera holds a secret code that can only be activated upon scanning an RFID card against what appears to be an electrical box outside a coffee shop.

BounceBounce caches are typically two-step, so once you open the box and activate the camera, you then must solve the puzzle by watching the flashes of four colored LEDs in order to open the keypad and obtain the logbook.

③ SALMON RUN

This multi-cache provides a partnership experience to educate visitors at a local fish hatchery. Each step is clearly marked to guide you around the location, answering questions to unlock each destination leading to a fish ladder magnetic puzzle! Guide the key through the fish ladder maze to open the final box and sign that logbook.

④ IT TAKES THREE

The City of Issaquah, Washington, requested a geocache at a historical landmark, with the caveat that it had to blend into the environment and nothing could be changed at the location. Upon inspection, I noticed three hollow poles behind this historic Shell station. This cache literally requires three to six individuals to open, and you may be forced to ask pedestrians on the nearby trail to lend a hand!

CHAD CHAMPION aka BounceBounce lives in Seattle, Washington, and loves to make puzzle boxes, portable escape room cases, and geocaches. Find him on YouTube @ GeocacheTalk and Instagram @bouncebounce8.

Make a Magic Blocks Gadget Cache

Written and photographed by Chad Champion / Gadget Academy

TIME REQUIRED: A Weekend

DIFFICULTY: Moderate

COST: $40–$80

MATERIALS

Magic Blocks Gadget Kit $80 from Gadget Academy, gadgetacademybuilds.com/product/magic-blocks/2. Or source the parts separately and DIY it. Kit contains:

» **Magic Blocks printed circuit boards (PCBs), v1.3 (4)** Can buy separately at Gadget Academy.
» **Resistors, 220Ω, ¼W (12)**
» **7-segment LED digit displays, common cathode (4)** such as Amazon B00EZBGUMC
» **DIP switches, 8-position (4)** B07DSBX4BK
» **Battery holders, 2xAA (4)** B07BXZQQVK
» **Magnetic reed switches (4)** B07VMYG7GF
» **Foam, approx. 6"×1"×1"**
» **Philips screws, 8mm (8)**
» **Torx screws, T9 size, 8mm (16)** to deter tampering; or use M3 socket cap screws
» **Custom 3D-printed cases (4)** Find our STLs at Gadget Academy, or create your own enclosures.
» **Batteries, AA (8)** not included. We recommend Energizer Ultimate Lithium for long life, but this cache will work well on alkaline batteries as well.

TOOLS

» **Soldering station and solder**
» **Side cutters**
» **Scissors**
» **Small Phillips screwdriver**
» **Torx screwdriver, T9 size** from any hardware store

The Magic Blocks gadget cache is a fun and simple "proximity and order" puzzle. It consists of four gadget blocks. Each will illuminate a single digit that you program into it — but only when all four blocks are placed side by side, in the correct order! Strategically located magnets and magnetic switches are used to sense the proximity of each block.

The four digits can be used as a code, or a combination to a lock, or for any other purpose you like in your geocache. You program each magic block with the digit of your choice. Our kit is fairly easy to assemble, requiring only basic skill in soldering. We've designed a circuit board and a custom 3D-printed case to make it quick and easy. Or you can source your own parts and DIY it!

Each magic block requires two AA batteries, which provide for a very long operating life. When not in play, the magic blocks must be stored in a manner that keeps their sides separated by 1½" or more, or they may activate while hiding in your cache, and drain the batteries.

Special thanks to Trycacheus, for the Magic Blocks idea and his blessing to share it with you!

BUILD YOUR MAGIC BLOCKS

We suggest you assemble all four PCBs at the same time, unless you're not very familiar with soldering — in which case you might want to fully assemble one magic block and make sure it's all working before building the other three.

If you're new to soldering, brush up with this *Make:* video (youtu.be/v4D_Rdp1uh8) and primer (makezine.com/projects/primer-soldering-and-desoldering) and Adafruit's handy reference card (makezine.com/go/adafruit-solder-card).

Let's get started!

ASSEMBLE THE PCBS

1. Decide on the four digits you wish to use (e.g., 3 1 5 9). Write each digit on the back of one board, to help you keep track as you assemble them (Figure Ⓐ).

2. You will solder either one, two, or three 220Ω current-limiting resistors on each PCB, depending on the digit you selected for that block. That way, all the digits will have the same brightness. Following **Table 1**, pick the correct quantity of resistors for your blocks.

TABLE 1

Digit Selected	Resistors to Use
1, 7	R1 only
2, 3, 4, 5	R1 and R2
0, 6, 8, 9	R1, R2, and R3

3. Bend the leads on the resistors, then insert them in the PCB in position R1, as well as R2 and R3 if needed (Figure Ⓑ). Polarity (which lead goes in which hole) does not matter for resistors.

Flip the PCB over and solder the resistor leads, then trim them just above the solder joint with your side cutters (Figure Ⓒ).

Repeat for the remaining three PCBs, each with their chosen digit and adjusted number of resistors.

4. Insert the DIP switch, with the ON label toward the center of the PCB (Figure Ⓓ).

Flip the PCB and prop the other end so the DIP switch stays flat against the board. Solder any one pin on the switch and confirm it still sits flat.

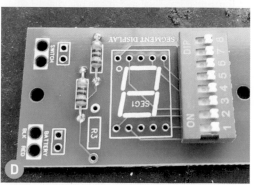

E

F

G

H

If needed, reflow the connection to make sure it's flat. Then solder the remaining leads and trim.

5. Insert the 7-segment display (Figure 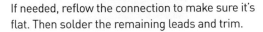E) with its decimal point at the bottom (when the DIP switch is on the left). As before, flip the PCB, keep the display flat, solder leads, and trim.

8. Insert the red and black battery leads from the top of the PCB in the holes indicated. Loop them on the back side, poke the tips through the other holes indicated, then solder *on the top*, and trim (Figure F).

9. Do the same for the magnetic sensing switch wires: Insert from the top, loop in the back, solder on top, and trim. This switch has no polarity, so use either wire in either position.

Tighten the wire loops on the backside. Looking good! The electronic portion of the build is complete (Figure G).

PROGRAM AND TEST

It's time to set your digits and turn them on! You'll program your selected digits in each magic block using the 8-position DIP switch. These tiny switches are moved to the ON position using a fingernail, pen tip, or small tool.

1. Following **Table 2** and the letter labels printed on the PCB, set your DIP switches to the ON position for the digit you selected on each magic block (Figure H).

TABLE 2

Selected Digit	Switches to Activate (PCB Labels)
0	A,B,C,D,E,F
1	B,C
2	A,B,D,E,G
3	A,B,C,D,G
4	B,C,F,G
5	A,C,D,F,G
6	A,C,D,E,F,G
7	A,B,C
8	A,B,C,D,E,F,G
9	A,B,C,D,F,G
Decimal point	dp

2. Insert two AA batteries in each battery holder, ensuring correct polarity.

3. Place a magnet close to each reed switch. The digit you selected for that magic block should illuminate, and then extinguish as you move the magnet away. Make sure all four assemblies are illuminating the correct segments (and digits) (Figure I).

That was (hopefully) quick and easy! Let's get all these assemblies mounted into their cases and let the magic begin.

FINAL ASSEMBLY

Time to put it all together! Note the direction arrow on the inside bottom of each case. For the following steps, orient all the cases with the arrow pointing up.

1. Mark the 6" (15cm) length of foam every 1½" (3.8cm) and cut it with scissors into four equal pieces (Figure J).

2. Select one of the PCB assemblies and center the battery pack in the bottom of the case, with the battery wires on the "up" side of the case (Figure K). Add a piece of foam on top of the battery, paper side up (Figure L). Then set the PCB on top as shown — DIP switch left, decimal point down — and affix in place with two screws (Figure M).

Repeat on the other three cases, and tuck the battery wires into the top of the case.

ENCODE THE MAGIC BLOCK ORDER

Now you can place the switches and activating magnets in such a way that all the blocks must be placed in order, side by side, to illuminate all the numbers simultaneously. This provides the geocache finder with your numbers in the correct order.

We'll show you one way to link the blocks, though other switch and magnet positions are possible.

1. Place all four blocks in a row, in the order you want to display the digits. (Check again with the magnet if you're unsure which is which.) For clarity, we'll call the blocks W, X, Y, and Z, from left to right (Figure N). As each block is set up, you'll move it off to the side, so pay careful attention to the W, X, Y, and Z designations at each step.

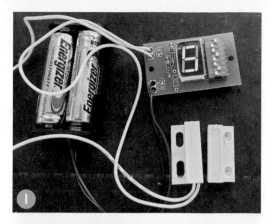

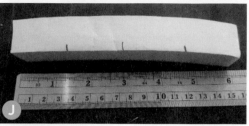

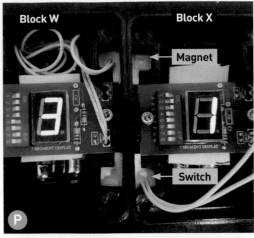

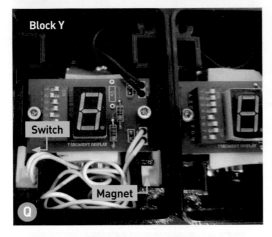

2. Block W — Slide the sensing switch from block W into either of the two positions on the right side of the block. Place a magnet in the remaining position on the right side of block W. Tuck in the switch wires in the top or bottom of the block (Figure O).

> **NOTE:** Depending which position you're using, the sensing switches may be placed wire side up, or they may need to placed wire side down, in which case you need to feed in the wire first. Don't force them down in the slot, but do insert the magnets and switches as low in the block as possible.

3. Block X — Slide the sensing switch from block X into the position (on the left side) that's opposite the magnet in block W. Also slide a magnet into block X, opposite the sensing switch in block W. Both blocks should now have their numbers

illuminate when next to each other (Figure P).

Set block W aside (no sense wasting your batteries). Add a second magnet to block X, in either position on the right side. Tuck in the switch wires in the top or bottom of the block.

4. Block Y — Slide the block Y sensing switch into the slot opposite the magnet in block X. Block Y will now illuminate when next to block X. Set block X aside.

Slide a magnet into the right side of block Y, in either position. Tuck in the switch wires (Figure Q).

5. Block Z — Slide the sensing switch from block Z into the position opposite the magnet in block Y. Block Z will illuminate its number when next to block Y. Move block Y away to extinguish the digit. Tuck in the switch wires (Figure R).

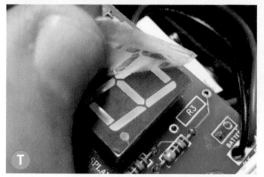

6. Confirm that when all four blocks are next to each other, in the correct order, they will all illuminate (Figure Ⓢ)!

Remove the protective film over the 7-segment displays (if present) and make sure the wires are tucked in so they won't block the view (Figure Ⓣ).

7. Remove the protective paper from the acrylic lens window inside of all four custom block covers (Figure Ⓤ).

Place the cover on all blocks, logo side down, making sure no wires are pinched under them, and attach each with four T9 Torx screws on each block.

The build is complete!

DEPLOY THE MAGIC!

Put the blocks next to each other in the correct order and confirm all digits are right side up and all the cover logos are aligned. Now you can put your Magic Blocks into your own geocache to mystify all who find it.

Don't forget, your cache design needs to keep the four Magic Blocks separated, until the cacher is ready to solve the puzzle! ◇

This huge clear dome came from a thrift-store Eero Aarnio-style Bubble Chair.

JOSHUA ELLINGSON is an illustrator, maker, and electronic and video artist living in San Francisco. ellingson.tv

Pepper's Ghost Hologram Flask

Create a spooky apparition in a jar, using any TV or mobile display

Written and illustrated by Joshua Ellingson

I learned about the classic stage illusion "Pepper's Ghost" from an episode of *Mr. Wizard's World* on Nickelodeon, sometime in the late 1980s. Many of his household science experiments were small and easy to replicate at home but occasionally Mr. Wizard would show off something really spectacular to his adolescent co-stars. In this particular segment, he produced an apparition of a skeleton onto a chair in a room that definitely did not have a skeleton in it before he cut the lights. He explained how the reflection illusion worked, with a large angled sheet of glass between the empty chair and the opened door. It was amazing, but I knew there was no hope of trying to create that effect at home. Also, where was I going to get a skeleton?

Thirty years later, in 2019, I was watching *The Imagineers*, a documentary on Disney+ about the history of Disney's Imagineering department and the development of the theme parks. The early episodes discuss Pepper's Ghost multiple times, especially in reference to the amazing usage of the effect in the Haunted Mansion attraction. It rekindled an old curiosity and now, as an adult, I realized that I probably had all the supplies needed to make my own Pepper's Ghost illusion.

So, I slid a piece of framing plexiglass at an angle into an aquarium, plopped an old LCD display on top of the tank facing down, and fired up a video of a goldfish on a black background. It worked great! I was amazed at how easy it was to make a convincing illusion of a swimming fish, so I continued with it. Eventually I tried out the effect with a bell jar on top of my iPad, with a rounded piece of polycarbonate as a reflector. This was the beginning of a whole new journey that led me through the worlds of photography, plastics, synthesizers, and real-time motion graphics.

Suddenly, after twenty-something years as a commercial artist, I was now a budding audio-visual installation artist? I had dabbled with video art but my previous experiments with black-and-white TVs now seemed to make more sense alongside the exploration of Pepper's Ghost.

It all felt like one big discovery and it all happened right before the pandemic hit. It turned out that I would soon end up having plenty of alone time to experiment.

Joshua Ellingson, Keith Hammond

TIME REQUIRED: 1–2 Hours
DIFFICULTY: Easy
COST: $20–$25

MATERIALS

» **Erlenmeyer flask, 1000ml** Karter Scientific 213G22, kartersci.com. Any size and brand will work but my template is for this specific flask.
» **Printed template** for the reflector. Download the free file *KarterTemplate1000ml.pdf* at makezine.com/go/peppers-ghost.
» **Cardstock paper** for a cutting template. An old cereal box or other food packaging would work well.
» **Polycarbonate sheet, 1/32" thick, 11×14** Lexan or similar. Any clear, thin, and flexible plastic will work but I like polycarbonate best; it's very flexible and thin sheets cut easily with scissors. Unlike acrylic, polycarbonate doesn't fracture when cut, so it can produce a clean edge. It's bendable enough to roll up and shove into a chemistry flask. I've also found most polycarbonate to be extremely transparent and excellent at catching a reflection.
» **A display of some type** I will be using an iPad, but a phone or upturned monitor would work as well.

TOOLS

» **Scissors**
» **Tape**
» **Permanent markers**
» **Long tweezers or needle-nose pliers (optional)** make it much easier to manipulate the reflector inside the flask

MY THING FOR TELEVISIONS

Just about any sort of display can pair well with a Pepper's Ghost, but I like **cathode ray tubes** the best. There's something mesmerizing about the glow of a CRT, especially black-and-white sets from the 1960s and 70s. They have an almost neon intensity that is impossible to reproduce with LCD and other more modern technologies. I have a growing collection of vintage televisions of various sizes, shapes, and ages. There's everything from a built-like-a-Buick 1958 RCA Sportable to a minuscule 1969 Symphonic Minni portable. I've got non-TV CRTs too, like a monitor that was originally used with a black-and-white video camera for microscopy. Mainly, though, I collect 1970s solid-state TVs from brands like Zenith and General Electric; their picture is often very strong, nearly 50 years on, and sometimes they're fashionable too. Case in point, I have a Zenith Sidekick that's covered completely in denim to match the blue jean craze of 1974.

I try to avoid opening the back of my TVs whenever possible, so I send video from my computer through a series of adapters that transform the digital signal to an analog format that's agreeable to my television's VHF antenna connection. HDMI-to-VHF all-in-one devices do exist but I haven't tried them out. They seem to be more expensive than a simple **HDMI-to-composite adapter/scaler** paired with an **RF modulator**. Remember RF modulators? Well, these days they seem to appear at every thrift store's electronics section for cheap.

Beyond the adapters, you need **cables** such as HDMI, composite video (RCA yellow plug), and copper coaxial, and if your TV only has screw terminal connectors you'll need a **matching transformer** with coaxial-to-spade connectors. If you're of a certain age, this can bring back memories of hooking up an Atari 2600. When everything is connected and working, it feels like magic to drag windows from the desktop over to the tube of the television.

SYNTHESIZERS, SOUND, AND VIDEO

Originally, I was posting my live-video experiments online using sounds from my record collection but I quickly realized I probably shouldn't be leaning too hard on other people's music for my own content. I started making original audio with a toy Stylophone, a handheld plastic thing that you control by touching or dragging a metal-tipped pen across a bed of metal keys. It sounds like a drunk chicken. Eventually, I graduated to an updated Stylophone with built-in reverb and some filter controls.

PEPPER'S GHOST ILLUSION — FROM DICKENS TO TUPAC

STAGE FRIGHT: The famed ghost projection technique was invented by English engineer Henry Dircks in 1858 in an effort to debunk spiritualists and to improve upon their "magic lantern" phantasmagoria shows. Debuted in the theater in 1862 by John Henry Pepper — the two men shared a patent — the illusion soon appeared onstage in *The Haunted Man* by permission of the author, Charles Dickens.

FATAL ATTRACTION: Board your Doom Buggy to be scare-tained by a gazillion ghosts in Disneyland's 54-year-old Haunted Mansion ride — most of them still created by animatronic puppets and the Dircks/Pepper illusion.

REANIMATED: Digital natives rediscovered Pepper's Ghost when the animated avatars of Gorillaz appeared onstage via 3D projection at music awards shows in 2005 and 2006, and again when the late Tupac Shakur performed "live" with Snoop Dogg and Dr. Dre at Coachella in 2012, kicking off a freaky fad for "hologram tours" by artists living, dead, and virtual. —*Keith Hammond*

Wikimedia Commons, Disney, Gorillaz, Keith Hammond

Those simple controls were my introduction to the concepts of subtractive synthesis, and I needed to learn more.

There are many options for beginner synthesizers and my first instrument ended up being the **Moog Mother-32**. I bought it because it seemed well organized and legible, even though I had no idea what anything meant yet. It has vintage looks, absolutely no screen, woodgrain sides, and red blinky lights. Part of my rationale was that even if I couldn't make it sound good, at least it would look great on camera. Over time, I acquired a **Moog Subharmonicon** and a **DFAM (Drummer From Another Mother)** to finish off the three-tier "Sound Studio" set. They each have their own focus and each opens up new possibilities in the others.

As it turns out, synthesizers are sort of addicting and it's easy to spend loads of time and money on this highly entertaining game of audio logic puzzles (see "Modular Synthesizers," *Make:* Volume 85). If you're interested in trying out the concepts of synthesizers without breaking the bank, I recommend **VCV Rack**. It's a sort of analog synthesizer emulation software that has a huge community of developers and users. VCV Rack can receive MIDI from hardware controllers and I use it all the time to output MIDI for controlling other things like video software, such as **VDMX**.

VDMX is a live-performance video software for Mac, and I use it to manipulate video that I send to my TVs. It's a peculiar piece of software and also very powerful. The UI seems to have remained in the late 1990s — there's still no Undo function, and windows float untethered to any sort of overarching framework. To the uninitiated, it might appear outdated, but there are plenty of reasons to push past that impulse, and I'm glad I stuck with it. VDMX can easily integrate system

audio and MIDI devices, it has simple options for spanning across multiple displays, and it's got a modular approach that allows for any slider or button to communicate any other slider or button inside of VDMX. It's miles to the left from Adobe and I deeply enjoy uncovering its mysteries.

I try to use videos that exist in the public domain for most of my projects. The **Prelinger Archive** at The Internet Archive has an excellent and extensive collection of videos in the public domain. Originally, I was just trying to find interesting footage that wouldn't content-match on YouTube but it didn't take long to become enamored with these time capsules of ephemeral film. There are vintage cigarette advertisements promoting the freshness of one brand over another, cringe-inducing etiquette reels from the 1950s, random cross-country amateur roadtrip films, and corporate promotional films about the latest advances in one technology or another. It's a deep well of moving images to pull from and also great for sound.

MY STUDIO SPACE

Like many people working in San Francisco, my workspace is also my living space. I rent a ground-floor studio apartment with plenty of storage. Admittedly, it's not the most practical situation for a vintage TV museum / maker space, but I'm working with what I have. There's a pantry off the kitchen that I've repurposed into an art space and it's currently where I package and ship my enamel pin designs. I use an Intel Mac Mini in the main room to operate VDMX since it seems to be most stable in the pre-Apple silicon world. In the pantry studio, I have a bare-bones M1 Mac Mini for editing and rendering my short social media clips.

I still sometimes draw and paint, so the studio-

Keith Hammond

pantry is also where I leave the mess that I make while doing that.

The living area of the main room is now home to a giant polycarbonate dome atop a 55" flat-screen TV for Pepper's Ghost purposes. Originally it was a designer bubble chair that I found among the secondhand couches at Community Thrift. I decided it would be better taking up my space next to my couch instead. My desk is a vintage Steelcase Tanker Desk that was handed down to me from a video editor friend. It had plenty of surface area which I have filled completely with computer and audio gear.

By necessity, the living area is also a photography studio with a few light stands and tripods placed here and there at all times. I shoot with a Canon M6 Mark II sideways on a tripod for capturing vertical video. I use a Ninja V recorder to get the most out of the HDMI feed from the camera. A diffusion filter in combination with an ND filter on the Sigma 16mm lens helps manage the wild brightness from the CRTs and LEDs. Sometimes, I'll use a vintage Russian lens with an adapter and speed booster to create wider or macro shots.

There is a lot of trial and error with filming Pepper's Ghost videos. Depending on the screen and the image itself, the environmental lighting needs to be adjusted accordingly. For the effect to be vibrant, it always needs to be a little dark. It's also important to see the environment around the effect, so it can require a balance of small lighting adjustments. I also use subtle color and lighting post-production adjustments in DaVinci Resolve video editing software to help re-create what the video looks like in-person.

MAKE A MAD SCIENTIST PEPPER'S GHOST FLASK

You don't need a giant bubble dome to create your own Pepper's Ghost illusion. It's a very scalable effect, requiring only simple materials. that are easy to come by. I've even made demo videos about how to make a Pepper's Ghost with a domed coffee lid, snack packaging, and your phone.

Since Halloween is always just around the corner, let's make a Pepper's Ghost Chemistry Flask. It will be right at home in a haunted house, mad scientist's laboratory, or Halloween party.

1. PRINT THE TEMPLATE PATTERN
Download the template file for the reflector from the project page at makezine.com/go/peppers-ghost, and cut it out (Figure Ⓐ). This shape will sit inside the flask at an angle to catch the reflection of the image on the display. The shape

Ⓐ

should match the geometry of the flask so that the plastic isn't very visible. To get it right, we'll need to practice on card stock.

2. CUT THE TEMPLATE
Trace the printed template onto the cardstock paper and cut it out. Cardstock is more rigid than

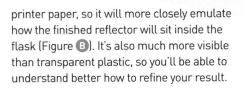

Joshua Ellingson

printer paper, so it will more closely emulate how the finished reflector will sit inside the flask (Figure **B**). It's also much more visible than transparent plastic, so you'll be able to understand better how to refine your result.

3. TEST-FIT IN THE FLASK

Roll up the cardstock cutout and shove it inside the flask. Move it around with tweezers or pliers to position it at something close to a 45° angle (Figure **C**). Take note of any bending or other issues with the angle. If the cutout isn't sitting evenly at an angle, remove it and either trim to fit or start over with a new piece of cardstock. It's worth it to try a few times if needed.

4. CUT THE REFLECTOR

Once you've arrived at an acceptable cutout, use it as a template for the clear polycarbonate plastic sheet. If your plastic has a thin protective overlay, leave that on while tracing your template with permanent marker (Figure **D**); that way, you can peel it away with your marks after the cut is complete. Carefully cut out the plastic along your trace (Figure **E**) with scissors.

5. ROLL IT UP

This plastic is very flexible; it should roll up like a

fruit-snack without creasing or breaking (Figure **F**).

6. SHOVE IT IN

Wide end first, shove your plastic roll-up into the flask (Figure **G**). If everything went correctly, the plastic reflector should pop open into the same position as the cutout template (Figure **H**).

7. CHOOSE A TEST VIDEO

Before settling on a final subject matter for your Mad Scientist Pepper's Ghost Chemistry Flask, it's a good idea to test out what you've made so far to ensure that the effect is working. Find a video or image on a black background. Anything black in a Pepper's Ghost illusion becomes transparent, so the focus of the video will appear to float.

A quick Google search will yield loads of stock video of various creatures and objects against a black background. Find one with bright colors against that very dark or black background, There's one in particular that I use so often that I paid for an HD version of it, and the goldfish has sort of become my mascot.

8. TEST YOUR PEPPER'S GHOST FLASK!

Dim the lights and place your prepared flask on top of the video screen (Figure **I**). It's alive!

You might notice that the image is upside down;

in that case, just rotate the display so that it's facing the other direction.

Also remember that anything in the flask will be a mirror image of what it is reflecting. This could be important if you're thinking of putting text inside the flask. If so, you might use a video editing software to flip the video horizontally so that it reflects correctly.

FLASK FICTION

Now that you have a working Mad Scientist Pepper's Ghost Chemistry Flask, all that's left is to elaborate on your creation. What would it look like filled with green liquid? Creatures, spirits, eyeballs, or body parts? Do you have other chemistry glass that could use a floating experiment inside? Would you like to make your own animation?

ANIMATION OPTIONS

If you're using an iPad, the popular drawing app Procreate has options for simple animation. It's easy to make a frame-by-frame animation against a black background and then loop it as an animation inside the flask (Figure **J**). If you're familiar with other motion graphics software like Adobe After Effects then there are loads of possibilities for custom mad scientist creations.

IPAD AS EXTERNAL DISPLAY

Sometimes, I use Apple's Display Preferences to turn my iPad into an external display for my Mac (a feature called Sidecar). Then I use real-time video software VDMX to produce audio-reactive Pepper's Ghost effects from my iPad's screen into a glass container. Although it has great MacOS integration, Sidecar can behave unexpectedly when running video software. A great alternative might be the YAM (Yet Another Monitor) app for Mac and iOS. YAM seems to have very low latency and supports iPhone and other devices.

ALL SHAPES AND SIZES

You can use almost any size or shape of glassware or other transparent containers for this project. I've used cheap snow globes, and bell jars ranging from gigantic to this tiny one (Figure **K**) that fits on my TinyTV (tinycircuits.com). ⊘

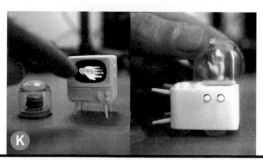

Luminary Geometry

ANURADHA REDDY is a researcher who combines crafts, culture, and electronics to shed new perspectives on hardware.

ARTURO182 builds open-source hardware and is the founder of the Swedish company Solder Party.

Written by Anuradha Reddy with arturo182

Period and lunar tracking with an illuminating calendar

This project manifested at the time of the overturning of the U.S. Supreme Court's *Roe v. Wade* decision in June 2022. As the news spread like wildfire, menstruating people worldwide were quickly advised to delete their period-tracking data from private apps and devices, to hide their period data from their doctors, and even their partners, to avoid future incrimination around the right to abortion.

Shocked and angry, I needed an outlet, and what better outlet than making?

ISLAMIC GEOMETRY

A month or so before the news hit, I had taken an online Islamic geometry course by Samira Mian (samiramian.uk). Soon after, I began experimenting with the patterns in code (p5js). The code got me thinking about translating the

arturo182

patterns into hardware, and I proposed this idea to arturo182. Inspired by Jason Coon's LED CycloHex patterns, we attempted an LED Islamic geometry-inspired circuit board. We chose the ten-pointed star from a pattern book I found in my late grandfather's dusty library on my last visit to India.

The 10-pointed star was an easy pattern to replicate and tessellate with digital software. Arturo182 designed the entire circuit board from scratch in KiCad. It incorporated addressable surface-mount LEDs — 10 for each star unit in a 6×5 grid (30 units = 250 LEDs, because some LEDs overlap by the nature of the geometric pattern). We consulted each other on the look and feel of the board: purple solder mask, white silkscreen, and gold for the traditional art of Islamic illumination. At the time, we aimed to experiment with NeoPixel libraries and play with LED animations. We wanted to have fun and make a pretty board, but had no idea about the new direction the project would take.

Soon the most beautiful Islamic geometry-inspired circuit board was in our hands. We hooked it up to Solder Party's RP2040 Stamp Round Carrier and used CircuitPython to program it. But I was far from satisfied with the blinky LED animation trains, which I thought took the focus away from the timeless artform on the circuit board. I wanted to make it appear delicate, drawing attention to the art.

MANY MOONS

Meanwhile, my obsession with Islamic geometry grew, and I learned all sorts of interesting meanings and symbolisms embodied in the artform — symmetry, unity, harmony, and balance. By drawing connections to symmetrical natural cycles sacred to many cultures and practices, I landed on experimenting with moon phases on the board. Arturo182 helped manually program each unit of 10 LEDs using a coded array to reflect four phases of the moon (first-quarter, full moon, third-quarter, new moon). A total grid of 30 ten-pointed stars allowed us to visualize the board as a monthly moon phase calendar! Another silver lining was learning that lunar months alternate between 29 and 30 days, so we could ignore months with 31 days.

TIME REQUIRED: A Weekend
DIFFICULTY: Intermediate
COST: $40–$50

MATERIALS
» **Printed circuit board** You can order the PCB from a service like OSHPark, using our KiCad files at github.com/arturo182/strange_leds_hw.
» **RGB LEDs, SMD 2×2mm package (250)** Worldsemi WS2812B-2020, e.g. part number C965555 from lcsc.com
» **Capacitors, 0.1µF, 0603 package (30)**
» **RP2040 Stamp microcontroller with RP2040 Stamp Round Carrier board** both from Solder Party, lectronz.com/products/rp2040-stamp-round-carrier

TOOLS
» **Soldering iron**
» **Solder**
» **Flux**
» **Computer with Circuit Python code editor** e.g. Mu Editor

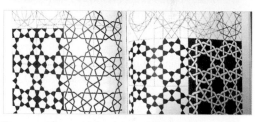

Two pages from an Islamic geometry pattern book show the process of making a 10-pointed star grid inside a rectangle.

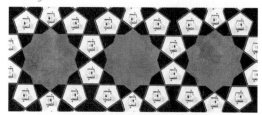

A close-up of the star pattern on the PCB. It has purple solder mask, white silkscreen, and gold for illumination.

A 3D-printed stand made with purple PLA holds the 2040 Stamp Round Carrier board, which drives the main LED board.

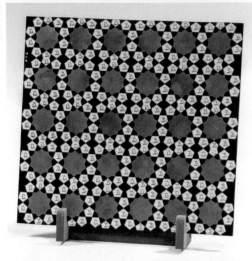

The completed PCB with LEDs arranged in an Islamic geometry pattern repeating in a 6×5 grid.

We initially chose white (**#FFFFFF**) for the LEDs representing the moon phases (the new moon was on a dimmer setting to indicate a fresh start). The board looked beautiful, but it still felt scant. It needed something more.

With the overturning of *Roe v. Wade*, it didn't take me long to imagine my monthly period tracking data alongside the moon phase calendar. This is when the project came together. I manually programmed the LEDs to reflect the two natural cycles in delectable twilight colors: **#330019** (deep rose) for period tracking, **#190033** (dark purple) for half and full moon phases, and **#0F000F** (dark magenta-purple) for the new moon. For this, I used an online tool (rapidtables.com) that includes an RBG color picker and converts a chosen color to a hex code. I experimented with a bunch of twilight hues on the LEDs before I came to a final decision. It made sense to bring more beauty to these natural cycles and to see them together — a way to push back against the misogyny surrounding our female bodies.

CIRCUITPYTHON CODE

The project code is rather short. Download *code. py* from the GitHub repo at github.com/anu1905/ Islamic-Geometry. Start by importing the **board** library for pin information and the **neopixel** library for controlling the LEDs:

```
import board
import neopixel
```

Then define the pixel pin on the board to control all 250 LEDs. If you're not using the 2040 Stamp Round Carrier, ensure that you have the right pixel pin for your board. Define the number of NeoPixels:

```
pixel_pin = board.GP11
num_pixels = 250
```

Next, write an array for a total of 30 star units (moons) of 10 LEDs each. Make sure to get the numbers right — remember, some LEDs overlap!

```
moons = [
[0,   1,   2,   3,   4,   5,   6,   7,
8,   9],
[10,  6,   7,   11,  12,  13,  14,
15,  16,  17],
[18,  14,  15,  19,  20,  21,  22,
23,  24,  25],
[26,  22,  23,  27,  28,  29,  30,
31,  32,  33],
[34,  30,  31,  35,  36,  37,  38,
39,  40,  41],
[42,  38,  39,  43,  44,  45,  46,
47,  48,  49],
[50,  51,  52,  53,  54,  55,  56,
57,  58,  59],
[60,  56,  57,  61,  62,  63,  64,
65,  66,  67],
[68,  64,  65,  69,  70,  71,  72,
73,  74,  75],
[76,  72,  73,  77,  78,  79,  80,
81,  82,  83],
[84,  80,  81,  85,  86,  87,  88,
89,  90,  91],
[92,  88,  89,  93,  94,  95,  96,
97,  98,  99],
[100, 101, 102, 103, 104, 105, 106,
107, 108, 109],
[110, 106, 107, 111, 112, 113, 114,
115, 116, 117],
[118, 114, 115, 119, 120, 121, 122,
123, 124, 125],
[126, 122, 123, 127, 128, 129, 130,
131, 132, 133],
```

```
[134, 130, 131, 135, 136, 137, 138,
139, 140, 141],
[142, 138, 139, 143, 144, 145, 146,
147, 148, 149],
[150, 151, 152, 153, 154, 155, 156,
157, 158, 159],
[160, 156, 157, 161, 162, 163, 164,
165, 166, 167],
[168, 164, 165, 169, 170, 171, 172,
173, 174, 175],
[176, 172, 173, 177, 178, 179, 180,
181, 182, 183],
[184, 180, 181, 185, 186, 187, 188,
189, 190, 191],
[192, 188, 189, 193, 194, 195, 196,
197, 198, 199],
[200, 201, 202, 203, 204, 205, 206,
207, 208, 209],
[210, 196, 197, 211, 212, 213, 214,
215, 216, 217],
[218, 214, 215, 219, 220, 221, 222,
223, 224, 225],
[226, 222, 223, 227, 228, 229, 230,
231, 232, 233],
[234, 230, 231, 235, 236, 237, 238,
239, 240, 241],
[242, 238, 239, 243, 244, 245, 246,
247, 248, 249]
]
```

Then define how the LEDs correspond with the moon phases and the period days. We define four moon phases in this project — full moon (10 LEDs), first-quarter (6 LEDs), third-quarter (6 LEDs), and the new moon (10 LEDs, second color). The period day is drawn the same as the full/new moon but we define it separately so we can later assign the right color to it.

```
def show_fullmoon(moon, color):
  for i in range(10):
    pixels[moons[moon][i]] = color
```

```
def show_firstquarter(moon, color):
  for i in range(5):
    pixels[moons[moon][i]] = color
  pixels[moons[moon][9]] = color
```

```
def show_thirdquarter(moon, color):
```

```
  for i in range(4, 10):
    pixels[moons[moon][i]] = color
```

```
def show_newmoon(moon, color):
  for i in range(10):
    pixels[moons[moon][i]] = color
```

```
def period(moon, color):
  for i in range(10):
    pixels[moons[moon][i]] = color
```

The following is most important part of the code because it's where you edit the moon phases and period days for every calendar month. For example, in the code below, period days fall between the third and seventh day of the month (i.e., define period units **2, 3, 4, 5, 6**), full moon on the eighth day, and so on. Moon phase information can be found on a website like timeanddate.com.

```
period(2, 0x330019)
period(3, 0x330019)
period(4, 0x330019)
period(5, 0x330019)
period(6, 0x330019)
show_fullmoon(7, 0x190033)
show_firstquarter(15, 0x190033)
show_newmoon(22, 0x0F000F)
show_thirdquarter(29, 0x190033)
```

PERIOD POWER

Seeing my period days overlap with the moon phases gave me a feeling for when my next cycle will arrive. Despite being a manual process of updating the board each month, I feel more empowered to visualize my cycle, locally and privately, rather than hide it from those who wish to take control of female bodies and their data. ●

EXTRA RESOURCES:

Programming NeoPixels in CircuitPython (for further play): learn.adafruit.com/circuitpython-essentials/circuitpython-neopixel.

RGB color picker to hex code conversion: rapidtables.com/web/color/RGB_Color.html

Weird Resistors Part 2
Squish & Spin!

Sew a squeezable resistive sensor, then dial up a soft potentiometer

Written and photographed by Lee Wilkins

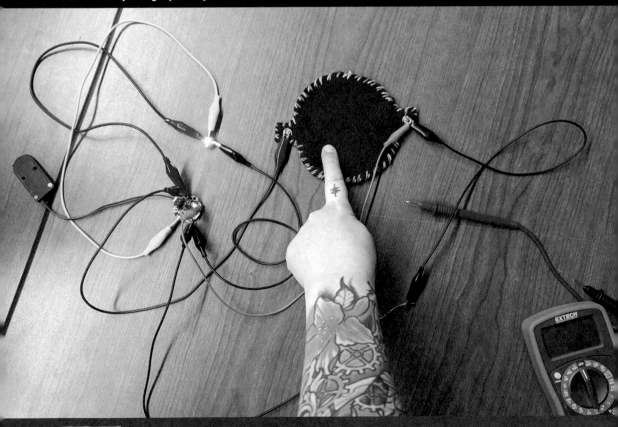

LEE WILKINS is an artist, cyborg, technologist, and educator based in Montreal, Quebec, a board member of the Open Source Hardware Association, and the author of this column on technology and the body and how they intertwine. Follow them on Instagram @leeborg_

In the previous installment of Weird Resistors ("Play With Your Food," *Make:* Volume 85), we talked about materials that have different resistive properties that might not be "traditional" resistors. This time I'd like to share with you some ways to make resistors more squishy and twisty. We'll make two kinds of weird resistors, and learn a bit more about where to find resistive materials. This is part of my upcoming zine on resistors!

FINDING RESISTIVE MATERIALS

We hear a lot about conductive fabric and thread, which can help build soft circuits, but materials with resistive properties can be useful too! Every material, even a copper wire, has some resistance, and you can find this spec when buying anything conductive; it's usually measured in **ohms per inch**, **ohms per foot**, or **ohms per meter**, depending on where in the world your manufacturer is located. Sometimes this can be tricky to get in order, so I like to use a unit converter when buying conductive and resistive materials to make sure I'm not misunderstanding the specs.

There are a few ways to make your own **variable resistors**. One way is to use materials that have a high resistance per inch, and to force a circuit through a specific amount of material to manipulate resistance. There are also some materials that change resistance when they are squished or strained; these are called **piezoresistive** materials. You can make these materials a part of your circuit and then activate their variable resistance by squishing, pressing, pulling, or twisting them!

A good place to start with variable resistor materials is with **Velostat**, sometimes called **Linqstat**. You can buy it on Adafruit or SparkFun, but it's also found in some garbage bags or other industrial materials. Another common piezoresistive material is **antistatic foam**, which is often shipped with electronics to protect them.

Some of my favorite fancier resistive materials come from **Eeonyx**, who make some great nonwoven heating textiles; the high resistance of these materials produces heat, but can also be used to make resistors. Check out EeonTex NW170-PI-20 (SparkFun COM-14110 or Adafruit

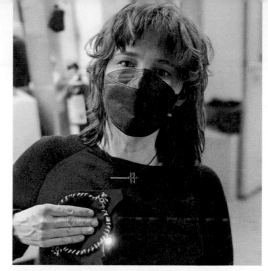

Marie LeBlanc Flanagan demonstrates a pressure sensor they made from fabric in the Soft & Squishy Sensing Switches workshop at the Milieux Institute for Arts, Culture, and Technology, at Concordia University in Montreal.

Alexandra Bachmeyer teaching a Soft Switches workshop at the Milieux Institute.

3670), or the stretchy EeonTex LTT-SLPA (Adafruit 3669). **Inntex** also makes a great resistive tape, and periodically stocks a resistive yarn.

While some products are specifically labeled resistive, you can achieve similar effects by using cheaper conductive materials that have a high resistance per meter. For most analog circuits, you'll need to make sure your DIY resistor is operating in the desired range. But if you're using a microcontroller, you might not need a huge range of resistance because you can translate your sensor input into an output in software.

Now let's make two types of sensors, Squish and Twist, and then look at how to test and calibrate them. This is part of a workshop I ran with Alex Bachmayer at the Milieux Institute.

SQUISH IT!
How to make a DIY pressure sensor

We're going to make a pressure sensor using Velostat that changes resistance when you squish it. Our sensor will have snaps that can be applied to any e-textile or wearable project! It can be used with or without a microcontroller.

This technique can be used to make all kinds of piezoresistive sensors that you can interact with in different ways — this is just the beginning!

1. CUT YOUR PIECES

Your sensor will be made out of a series of layers: two outer layers that are insulator fabrics (felt) holding it all together (shown in cyan in Figure Ⓐ), two pieces of conductive fabric (gray), and one piece of Velostat (black) that separates the conductive fabrics. In this circuit, the piezoresistive Velostat material stops the circuit from being completed, but as pressure is applied, its resistance drops, allowing current to flow.

You can cut your sensor in any shape you want! The rules are:

- Your entire circuit should be held together by two pieces of felt. These are the biggest — the bread of the sandwich, if you will (Figure Ⓑ).
- Your circuit should have two snaps, one on each end.
- Your conductive fabrics need to touch the snaps.
- Your conductive fabrics need to *not* touch each other, so your Velostat should be larger than the conductive pieces.

I recommend making your felt pieces at least 3" across, to give yourself enough space to work. You may also want to test beforehand to make sure your snaps are conductive.

Here are some examples of shapes I like. You can download my templates at makezine.com/go/squishy-sensors. Feel free to use them!

TIME REQUIRED: 1–2 Hours

DIFFICULTY: Easy

COST: $10–$30

MATERIALS
- » Felt
- » Velostat conductive sheet
- » Iron-on conductive fabric
- » 4-piece metal snaps (2)
- » Thread, non-conductive

TOOLS
- » Scissors
- » Needle
- » Snap applicator
- » Multimeter (optional)
- » Gemma microcontroller (optional)

Ⓐ

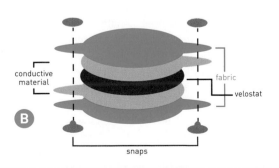

Ⓑ

conductive material

fabric

velostat

snaps

2. SEW YOUR SANDWICH

Once your pieces are cut out, make sure they all fit together. It helps to stack them up and be sure all the rules are being followed. Then, iron your conductive fabric onto the felt, place the Velostat in the middle, and sew your entire circuit shut as indicated (Figure C and D).

Finally, apply your snaps with the snap applicator. Make sure the snaps touch the conductive fabric!

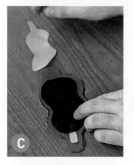

TWIST IT!
How to make a DIY soft potentiometer

Next we'll make a potentiometer out of resistive fabric. Then you can use this soft potentiometer to control your electronic projects with a twisting motion.

You can use this same technique to make all kinds of potentiometer circuits that you can slide, twist, or spin!

1. CUT AND ARRANGE PARTS

To make this circuit, we want to force the flow of electricity to go through different amounts of resistive material to create different resistance readings. For example, if our material has 10kΩ per inch resistance, we can use 5" of material for a difference of 50kΩ, depending where the connection is being made (Figure E).

For our circuit, we'll cut the template shown in Figure F using Eeonyx heating fabric as our resistive material, but you should feel free to use any shape that follows this concept. You'll place a snap at the center with a wire that can spin around the center point, making contact with the resistive material at different points.

TIME REQUIRED: 1–2 Hours

DIFFICULTY: Easy

COST: $10–$30

MATERIALS
» **Felt**
» **Resistive fabric** such as Eeonyx NW170-PI-20
» **Conductive thread or tape**
» **Wire**

TOOLS
» **Scissors**
» **Multimeter (optional)**
» **Gemma microcontroller (optional)**

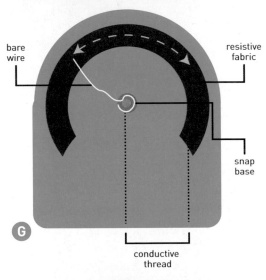

bare wire

resistive fabric

snap base

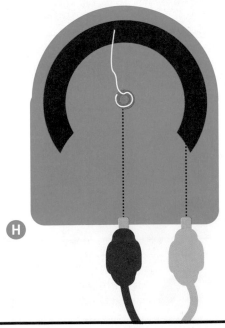

conductive thread

2. SEW, IRON, AND WIRE

Now sew or iron on the conductive thread or tape, sew down the resistive fabric, and make the rest of your connections as indicated in Figures **G** and **H**. Make sure the center wire — the "wiper" of the potentiometer — is long enough to contact the resistive fabric, and also loose enough to spin around the snap.

TEST AND/OR CODE YOUR SOFT SENSORS

Every DIY sensor is different, depending on how much resistive material you use and the properties of all your other materials. A good place to start is to take your multimeter and read the resistance. Place the leads on each end of your sensor and activate the sensor by pressing down, or by moving the wiper wire. If you don't see any change or very little resistance, your problem might be in your sensor construction. Check to see that the conductive pieces are separated by a resistive material.

To make a basic circuit, you can hook up your resistor to an LED and a battery as shown in Figure **I**, with wire or conductive thread.

If you want to get fancy, you can use a microcontroller to create effects. I like to use the Flora or Gemma microcontroller (Figure **J**) for wearables projects because they are small, flat, and easy to integrate into soft projects.

Because all DIY circuits are different, you can use a *calibration* function in your code to determine the minimum and maximum sensor values. The Arduino code below takes the first 5 seconds when the program starts and looks for the **sensorMin** and **sensorMax** values. Once the

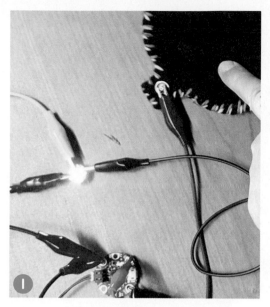

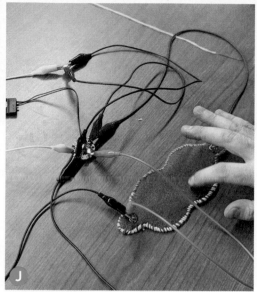

circuit turns on, press the sensor down a few times while it calibrates. I usually put this in my **setup** function:

```
int sensorPin = 0;
int sensorMax, sensorMin;
[...]
void calibrate() {
  while (millis() < 5000) { // Wait 5
seconds (5000 milliseconds)
    sensorValue = analogRead(0); //
Read your analog value
    if (sensorValue > sensorMax) { //
If it is greater than the previous max
value, it is the new highest value
      sensorMax = sensorValue;
    }
    if (sensorValue < sensorMin) { //
If it is less than the previous min
value, it is the new lowest value
      sensorMin = sensorValue;
    }
  }
}
```

You can then take your **_reading_** using a function like the following code, which reads your analog sensor value and returns a numerical value between 0 and 255:

```
int reading() {
  sensorValue = analogRead(sensorPin);
// Get the reading
  sensorValue = constrain(sensorValue,
sensorMin, sensorMax); // Constrain it
to be within your min and max values
  sensorValue = map(sensorValue,
sensorMin, sensorMax, 0, 255); // Remap
it to between 0 and 255
  return sensorValue;
}
```

This is great for getting the full range of color out of an LED, for example.

If you're using a modern Arduino-based microcontroller, you should use **pinMode INPUT_ PULLDOWN** which ensures that the value of the pin is controlled when it is not connected. This will allow you much more control over your circuit. If your microcontroller doesn't have that option, use a pulldown resistor (10kΩ between input and ground) for the same effect.

I have a few code examples at github.com/ LeeCyborg/SquishyGemma that I use when I run DIY resistor workshops; feel free to use, adapt, and share! Happy resisting! ✪

Want more? Learn all about switches, wires, and LEDs in my series of zines: leecyb.org/?page_id=383

Open-Source Incident Light Meter

Build a must-have photography tool and save hundreds!

Written by Martin Spendiff

MARTIN SPENDIFF is an easily distracted photographer living in Basel, Switzerland, with a few decades of experience in figuring things out. veeb.ch

Here's a Raspberry Pi Pico project that will spare you some tears when your films come back from the laboratory. We made an open source incident light meter, called Photon.

A light meter can be an essential tool in photography, especially with old film cameras. Modern cameras devote lots of computation to guessing how much light is falling on the subject. But if you can get to the subject and take a reading, no guessing is required and everything is easier! You can learn more at youtu.be/xju3yHBka7Q.

Photon reproduces some of the functionality of more expensive tools (Figure Ⓐ) using a few inexpensive, readily available parts. It measures ambient light brightness, as well as the red, green, and blue components of the light, which might allow white balance readings in future iterations. You could pay $200, or $2,000 — or solder your own for less than 50 bucks! It's easy.

BUILD YOUR PHOTON LIGHT METER
1. HARDWARE
Solder the power Shim to the Pico, following the Pimoroni instructions. Connect the battery.

Then connect the light sensor, OLED display, rotary encoder, and two switches to the Pico's GPIO pins (Figure Ⓑ) following the tables at

TIME REQUIRED: A Weekend

DIFFICULTY: Moderate

COST: $40–$50

MATERIALS
» **Raspberry Pi Pico microcontroller**
» **LiPo Shim battery charging board** Pimoroni PIM557, shop.pimoroni.com. Or use their Pico LiPo microcontroller, so you don't need the Shim.
» **OLED screen, full color, 128×128 pixels** Waveshare 1.5" module, Amazon B07DB5YFGW
» **Rotary encoder** to adjust settings and change priority mode; Iduino SE055 or similar
» **Momentary keyboard switch** to measure light; your choice of linear, tactile, or clicky
» **Momentary microswitch, 6×6mm** for ISO mode
» **BH1745 light/color sensor breakout board** Pimoroni PIM375
» **LiPo/Li-ion battery, 3.7V, 1100mAh** with JST plug
» **Enclosure** You can 3D print ours or design your own.

TOOLS
» **Soldering iron and solder**
» **3D printer (optional)**
» **Wire cutters/strippers**

1/60 second. ISO 400 =

A

B

Vanessa Bradley

github.com/veebch/photon. VCC and GND are also connected to the Pico, for all but the switches.

2. SOFTWARE
For the Pico firmware, download and install the UF2 image from github.com/pimoroni/pimoroni-pico/releases. (You need it to use the Pimoroni drivers for the light sensor.)

We wrote the Photon code based on simple math that converts the *illuminance* returned by the light sensor to *exposure value* (github.com/veebch/photon/#appendix). Clone the repo:

```
cd ~
git clone https://github.com/veebch/
photon.git
cd photon
```

Check the Pico port:
```
python -m serial.tools.list_ports
```

Using the port path (in our case **/dev/ttyACM0**), copy the code to the Pico using **ampy** (pypi.org/project/adafruit-ampy) and the **put** commands:
```
ampy -p /dev/ttyACM0 put drivers/
ampy -p /dev/ttyACM0 put gui/
ampy -p /dev/ttyACM0 put color_setup.py
ampy -p /dev/ttyACM0 put main.py
```

Done! Now the Python script will autorun.

3. ENCLOSURE
You can 3D print our STL files from the */cases* directory at github.com/veebch/photon.

USE IT
Press the Shim button to power up. Twist the rotary knob to choose aperture or shutter speed. Press the key to take the light measurement. For ISO mode, press the 6×6 switch, twist the knob to adjust ISO, then press the switch again to set it.

CONTRIBUTE TO THE CODE
If you can make Photon better, please fork the GitHub repository and use a feature branch. White balance readings should be doable! The most requested tweak is a flash/strobe mode — can you suggest changes that will allow such a fast measurement? We hope we've primed the pump to make Photon an open community resource. ⊘

Written and photographed by Forrest M. Mims III

Little Chip,
Big Gain

A mini course in op amps — the magical key to the analog world

While we live in the digital era, analog electronics still play a vital role. Analog circuits transform varying voltages and current into signals that digital systems can access and process. Power supplies, which transform wall power into the voltage and current required by many digital systems, are also analog in nature.

Transistors are key to both analog and digital circuits. An ancient junction transistor can be used as an analog amplifier, and it can function in a digital role simply by switching it on or off.

A transistor requires several outboard components to enable it to perform practical analog and digital functions. Those components have been combined on tiny silicon chips installed in dual-inline packages (DIPs) or much smaller surface mount devices (SMDs). The result is an enormous variety of digital logic chips and the highly versatile *operational amplifiers* popularly known as *op amps*.

Op amps have a very long history. Several engineers experimented with vacuum tube feedback amplifiers in the 1920s, and Bell Labs scientist Harold S. Black applied for a patent on the method. The patent office was not convinced the method would work, so they sat on Black's application for nearly a decade. Finally, in 1932 Black received US Patent US2102671 for his invention. Back then, op amps were made with vacuum tubes and were eventually used in analog computers for both military and scientific applications.

Op amps have inverting (–) and non-inverting (+) inputs (Figure **A**). The amplification or *gain* of an op amp is controlled by the negative feedback that occurs when some of the output voltage is coupled back to its inverting input by a resistor. The gain of the op amp is the *feedback resistance* (RF) divided by the resistance of the input resistor (R1). If RF is 100,000 ohms and R1 is 1,000 ohms, the circuit's gain is 100,000/1,000, or 100. If there is no input resistor, the gain equals the magnitude of the feedback resistance.

POPULAR OP AMPS

In 1968, Fairchild Semiconductor engineer David Fullagar designed the µA741 operational amplifier. The 741 was a breakthrough, for it greatly simplified the design of many kinds of electronic circuits. While the 741 initially sold for $70, Fairchild sold so many that price quickly dropped. 741s are still widely used, and a quick web search revealed a price range today of $0.70 to $19.36 for various versions of this 60-year-old chip.

When I began writing books for Radio Shack, the 741 was the only op amp they sold. The price was $0.79 (around $2.10 in today's money). Eventually dozens of op amps were developed with greatly improved performance over the pioneering 741. One of my favorites was and remains the Texas Instruments TLC271 family. These low-noise op amps consume very little current. While surface-mount versions of op amps dominate the market, the most popular versions for experimenters are easily used 8-pin DIPs (Figure **B**).

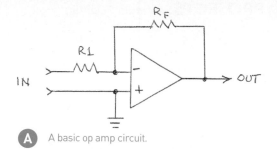

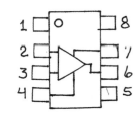

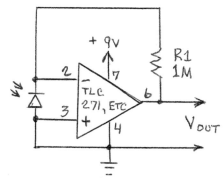

A A basic op amp circuit.

B This is the most common pin diagram for op amps installed in 8-pin DIPs.

C This basic op amp circuit amplifies the photocurrent from a small solar cell or photodiode a million times.

OP AMP LIGHT SENSOR

The simple circuit in Figure **C** illustrates the simplicity of a circuit that replaces two or three transistors and half a dozen resistors with an op amp. It's an ultra-simple light detector in which an op amp amplifies the photocurrent produced when a silicon solar cell or photodiode is exposed to light. Since there is no input resistor, the sensitivity of the circuit is controlled solely by feedback resistor R1. Thus, if R1 is 1 megohm, the photocurrent will be amplified 1,000,000 times, and very low levels of light can be detected.

While many different op amps can be used in this circuit, I often use the TLC271 op amp for ultra-sensitive light detection. When I designed the ultra-high gain twilight photometer described in these pages in 2015, I used a feedback resistor

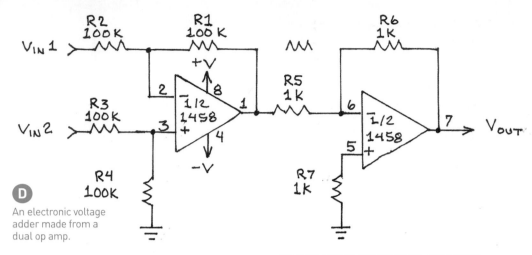

D An electronic voltage adder made from a dual op amp.

of two 40 gigohm resistors in series to provide an amplification of 40 or 80 billion! Thanks to this ultra-high gain, my home-built twilight photometers reliably detected aerosol layers up to 50km high.

This experience was a reminder that op amps are very flexible and that even inexpensive ones can perform well in a wide variety of applications. The circuits that follow illustrate this.

AN OP AMP ADDER

Though I used potentiometers to add voltages back in middle school days, the op amp adder I built for *Engineer's Mini-Notebook: Op Amp IC Circuits* (Radio Shack) made me fully realize the improved capabilities of op amp adders in analog computers. That adder is shown in Figure **D**.

The LM1458 dual op amp is one of many that can be used to make analog adders. In operation, two positive voltages are applied to the adder's two inputs. The negative sum of those voltages then appears at the output of the first op amp. The negative sum is then applied to the input of the second op amp, which inverts the negative voltage from the first to provide a positive output.

You can easily add more inputs to the basic circuit, each through a 10K resistor. The input voltages should be selected so that their sum is a few volts less than the power supply voltages.

This op amp adder illustrates a highly significant advantage of basic analog circuits over their digital equivalents. While a digital logic adder can provide much higher accuracy, the analog circuit is much faster.

IR REMOTE CONTROL TESTER

Have you ever pressed a button on a TV near-infrared remote controller and nothing happened? Figure **E** is a simple op amp circuit that will allow you to quickly determine if an IR remote is working.

In operation, a small silicon solar cell or photodiode is connected to the non-inverting input of an op amp through C1, a capacitor that blocks steady light while passing pulses of near IR from a working remote controller. R1 is a 100K resistor that amplifies the signal from the solar cell 100,000 times. If your IR remote control is working, the piezo element will emit an audible tone that matches the frequency of the IR signal from the controller. Be sure to use a piezo element, not a piezo buzzer.

You can connect an LED to the output of the op amp through a 2.2K resistor. The LED will flash at the same frequency as the controller and appear to be continuously on.

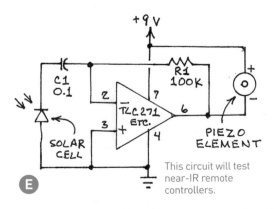

E This circuit will test near-IR remote controllers.

LIGHT-ACTIVATED ALARM CLOCK

Would you like an optical alarm clock triggered by sunlight? If you are self-employed or retired and want to rise with the sun, the circuit in Figure **F** is for you. The circuit is silent when the cadmium sulfide photoresistor is dark. When sunlight (or any other light source) strikes the photoresistor, the op amp switches Q1 on, which actuates a piezo buzzer.

Place the circuit near your bed so the buzzer will sound when light through a window is bright enough to begin your day. Potentiometer R1 can be used to adjust the sensitivity of the circuit.

As you have probably noticed by now, this simple circuit has other applications. For example, it can trigger a tone or alarm when a person or vehicle passes through a beam of light from an LED pointed at the photoresistor. Cadmium sulfide photoresistors respond best to green light, so you'll need to use a green LED for applications like this.

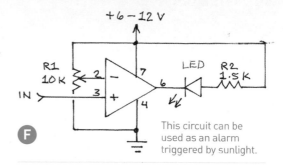

This circuit can be used as an alarm triggered by sunlight.

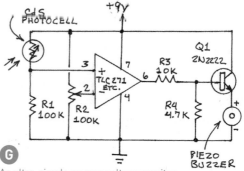

An ultra-simple op amp voltage monitor.

VOLTAGE MONITOR

Getting Started in Electronics, my best-selling Radio Shack book, concludes with 100 digital and analog circuits. One of the simplest analog circuits uses a 741 op amp (or any other popular op amp) to light an LED when an input voltage exceeds a level dialed in by a potentiometer. The circuit is shown in Figure **G**.

In operation, the LED is normally on. When the input voltage reaches a level determined by the adjustment of R2, which is connected as a voltage divider, the LED switches off. This operation can be reversed simply by switching the connections to input pins 2 and 3 of the op amp.

This very simple circuit has various practical applications. For example, it can be used to monitor the voltage from a battery that powers an electronic circuit. So long as the op amp receives power, R2 can be adjusted to inform a user when the voltage of either the op amp supply or an external supply falls below a level determined by the adjustment of R2.

When using a potentiometer, keep in mind that the resistance between the rotating (center) terminal and the terminal to which the pot has been most closely adjusted can be very low. If there's any chance that this low resistance

might damage a circuit, insert a resistor of a few thousand ohms between one or both opposing terminals of the pot and the circuitry to which they are connected.

GOING FURTHER

You may have noticed that the circuits in Figures F and G do not include a feedback resistor. That's because in these applications the op amp is connected as a *comparator*, a very useful circuit that illustrates how an op amp can be used in a digital mode.

The web is loaded with information about op amps. Many books about op amps are also available, including Walter G. Jung's classic *IC Op-Amp Cookbook* (search his name at Amazon Books). Another good op amp book is *Operational Amplifier Circuits* by John C.C. Nelson. Most of my Radio Shack books include op amp circuits, some of which are described in this column (search my name at Amazon books). ◗

FORREST M. MIMS III is an amateur scientist and Rolex Award winner, and was named by *Discover* magazine as one of the "50 Best Brains in Science." He has measured sunlight and the atmosphere since 1988. forrestmims.org

Complex Barrel Cams

How to design and 3D print hybrid cams for X-Y drawing and writing machines

Written and photographed by David Dalley

DAVID DALLEY (DaveMakesStuff) doesn't usually know what he's making until it's too late. He lives in British Columbia, Canada. linktr.ee/davemakesstuff

Are you looking for a way to engineer complex movements into your project? Let me introduce you to my friend, Cam.

WHO IS CAM?

Most likely, the *cam* mechanism closest to you right now is in your car. The valves in an internal combustion engine are controlled by cams. When the offset cams are rotated by the camshaft, they push the valves up and down in precise and coordinated movements.

Cams are often used in conjunction with *followers*. In Figure , the part of the valve or valve stem that rests on the cam is the follower — it follows the movement of the rotating cam. In this case, the rotational movement of the cam is translated into the linear movement of the valve.

This is an example of a simple cam, but don't let that fool you! Cam can be a pretty sophisticated fellow once you get to know him.

3D-PRINTED WRITING MACHINE

I had always wanted to design a 3D-printed writing machine. I knew that the constraints of 3D printing meant that I would have to minimize the number of moving parts and overall complexity of the machine. I knew I would have to find a way to encode and retrieve the text as simply and as elegantly as possible. Deep down inside, I knew that Cam had the answer.

CAM YOU DRAW A CIRCLE?

I started by trying to make a cam mechanism that could draw simple shapes, like a circle.

I studied *radial cams.* Like the example above of the valves in an internal combustion engine, radial cams translate the circumferential geometry of the cam shape into linear movement outward from the center of rotational axis of the cam. But this one-dimensional movement would not be enough to create two-dimensional shapes or text.

I studied *barrel cams,* aka *cylindrical cams.* In a barrel cam mechanism, the rotation of the cam translates into linear movement parallel to the rotational axis (Figure 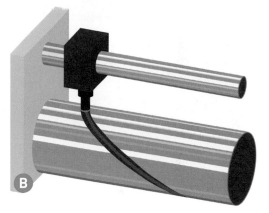). Again, this one-dimensional movement would not be enough to create two-dimensional shapes or text.

But what if I combined the two cam

mechanisms to create some sort of radial-barrel cam hybrid? Would that let me combine the two one-dimensional movements to create a two-dimensional path? I'd never seen it done before, but I decided that if I could figure out how to do it, I would call it a *complex barrel cam.*

CAM LOVES CAD

I opened up Onshape, my CAD program of choice. I started by creating a circle and drawing a horizontal line beneath it. I added eight reference points equally spaced around the circle; imagine them as points 1, 2, 3, etc. Then I created eight rotational instances of the circle and its eight reference points, using the line as the axis of rotation. Imagine them as instance 1, 2, 3, etc.

So now I had the eight points of a circle appearing eight times around a rotational axis, as in the second image in the sequence on the following page.

I opened up the Spline tool and that's when the magic happened! I created a closed spline by connecting point 1 on instance 1 with point 2 on instance 2, then point 3 on instance 3, and so on, until the spline created a complete wobbly-looking circle around the axis of rotation.

Text with reference points.

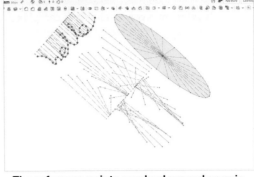

The reference points revolved around an axis.

The cam path plotted on the reference points.

Building a groove for a ball-shaped follower.

Building more of the cam structure.

The complete cam structure.

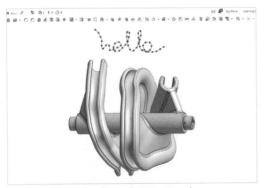

Complex barrel cam, top view.

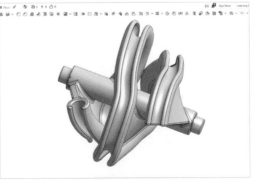

Complex barrel cam, front view.

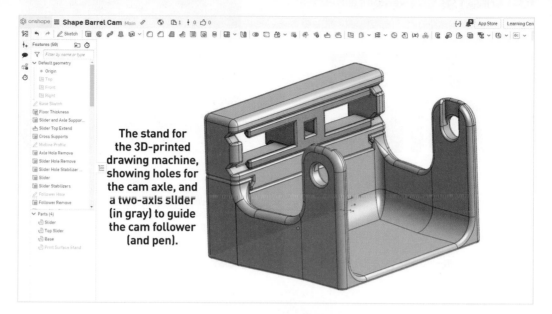

The stand for the 3D-printed drawing machine, showing holes for the cam axle, and a two-axis slider (in gray) to guide the cam follower (and pen).

This wobbly-looking spline had encoded the Cartesian X-Y coordinates of the circle on the plane into polar coordinates around an axis of rotation. The Y dimensions of the circle were now encoded as deviations in radial distance from the axis, and the X dimensions of the circle were now encoded as movement back and forth parallel to the axis.

DOING THE BEST I CAM

Once I had the magic spline, the rest was just hours and hours of tedious CAD work! I encircled the spline with a track that would hold and guide the ball-shaped end of a follower. I added structural elements to allow the track to rotate around its original axis of rotation. Then I created a stand, follower, and guides to hold it all together.

Many iterations later, I had a working model! The cam had encoded the Cartesian coordinates of the circle into polar coordinates around an axis, and by turning the cam, the polar coordinates were translated back out, through the follower, into the Cartesian coordinates of a circle on a plane.

Dimensional accuracy is critical for a project like this. The models pictured were printed on my Lulzbot TAZ Pro 3D printer.

BUT CAM IT WRITE?

Going from drawing shapes to writing text is just a matter of adding scale and tedium. Instead of eight points around a circle, an eight-character line of text requires 200+ reference points! The 200+ reference points are then multiplied and rotated around the axis 200+ times! Within those thousands of reference points lies the elegant path of a spline that can then be used to construct your complex barrel cam writing machine.

I THINK I CAM, I THINK I CAM

So, is encoding text on a complex barrel cam an efficient way to communicate? Not at all! It would take months to write a single sentence. Is it an intriguing way to demonstrate the power and potential of cams? Absolutely.

If you were to design a similar project, what would your Cam say? ◐

Build instructions and free downloads from Thangs:
- Drawing Machine: than.gs/m/332945
- Writing Machine: than.gs/m/42309

Videos of complex barrel cams in action:
- Drawing Machine: youtu.be/DdSJuxJodQM
- Writing Machine: youtu.be/-QQxDgqsT68
- Cam I Am: youtu.be/7daG2EHLoz0

String Walkers

A tiny toy that's big fun — and a medium challenge to make!

Written and photographed by Bob Knetzger

TIME REQUIRED: A Few Hours
DIFFICULTY: Moderate
COST: A Few Dollars

MATERIALS
» Acrylic sheet, ⅛" thick
» Paper clip, small
» Thread
» Miniature washers

TOOLS
» Laser cutter
» Snips
» Acrylic solvent
» Sandpaper
» Cyanoacrylate (CA) glue

BOB KNETZGER is a designer/inventor/musician whose award-winning toys have been featured on *The Tonight Show*, *Nightline*, and *Good Morning America*. He is the author of *Make: Fun!*, available at makershed.com and fine bookstores.

For the 50th installment of this column, here's a project that is just like "Toy Inventor's Notebook" itself: silly, fun, and with a bit of interesting science, all in a tiny package: string walker toys!

WALK LIKE AN EGYPTIAN (CAMEL)

These little toys work by gravity power. Hang the tiny weight over the edge of a table and the toy waddles and walks along — and then magically stops right at the edge. Over the years there have been different versions. a pair of lunar landing space men (Figures A and B), a Disney-licensed plodding Pluto, Jiminy Cricket wheeling a string bass, a shuffling buffalo, and many more (Figure C).

The most iconic version is the waddling camel. Just like the real thing (Figure D), this little toy camel has a *pacing* gait, where *ipsilateral* limbs contact the ground in unison. That means it walks with a rocking motion, with the front and rear legs on the same side moving forward at the same time.

HOW DOES IT WORK?

The potential energy of the hanging weight is converted into the kinetic energy of the mechanical walking action. But instead of the weight instantly dropping and quickly dragging the toy across the table, the toy's walking action acts like a mechanical escapement. It takes time for the camel to slightly tip to one side, swing its legs forward, sway back to the center, and then tip to the other side, alternately moving the other set of legs. The back and forth, pendulum-like motion s-l-o-w-s the action down as well as creating the cute, camel-walk effect.

Why does it stop instead of falling over the edge? At that point there is no longer any horizontal vector pulling the toy forward (exploratorium.edu/snacks/vector-toy).

MAKE YOUR OWN GRAVITY STRING WALKER

Go online at makezine.com/go/gravity-walker to download the SVG file for your laser cutter. I used a Glowforge with Proofgrade Medium acrylic for great results, but you can use any ⅛" acrylic material and laser cutter to cut the parts (Figure E).

The legs need to swing freely, so I laser cut

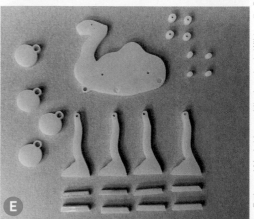

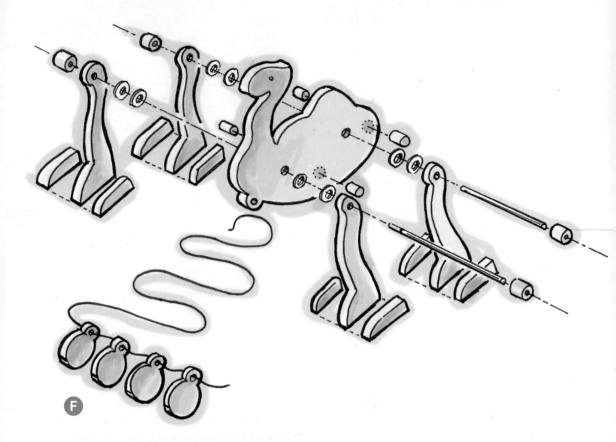

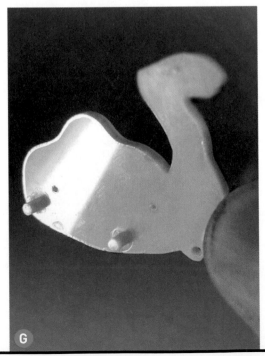

holes for steel axles snipped from a small size paper clip. I used the default etch settings for the two circle locators on the body and a much deeper etch to make half-thickness foot parts. Adjust for your laser cutter's kerf and power/speed etch settings.

THE DEVIL'S IN THE DETAILS

As simple as this toy looks, there are a few important design details that create the walking action (Figure **F**). Small pegs act as stops to limit the legs' backward travel. There are two etched circles that show you where to solvent bond the pegs (Figure **G**). Then, using those pegs as a position guide, bond a second set of pegs on the other side of the camel.

The feet need to be made wider and heavier. Add the small parts to both sides of the legs with solvent. Note that some of the feet parts are deeply etched to make them thinner. Those thin parts go on the inside of the legs. The thick

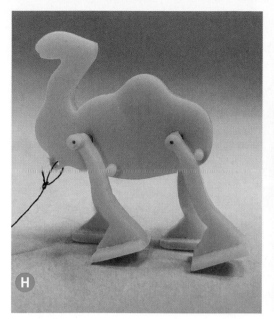

parts go on the outside. This gives clearance for the feet to swing past each other without hitting (Figure **H**), like the original toy (Figure **I**).

If needed, make more clearance by inserting small spacing washers on the axle between the legs and the body. Attach the small acrylic friction-fit collars on the ends of the axles and adjust the spacing for easy leg action. Then lock the collars in position with a tiny drop of cyanoacrylate glue on the very tip. Tie three or four of the weights together on a piece of thread and tie the other end to the hole on the front of the camel.

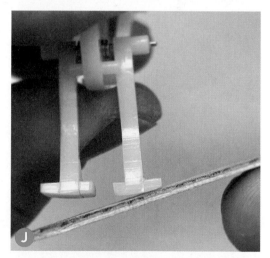

Lastly, create a slight bevel on the bottom of the feet. This allows the camel to sway from side to side. Use a small sanding block or emery board to carefully sand down the outside of the feet (Figure **J**). Remove a little material, test the tipping/walking action, and remove more material as needed.

It may take a bit of experimentation to get the best walking action (it did for me) but you'll have a real appreciation of the cleverness and precision of these tiny, inexpensive toys!

Of course you can also create your own original design. How about a *Star Wars* mini AT-AT? Here's a design for a cute Imperial ramp walker with a bunny-like hopping action: instructables.com/ Imperial-Ramp-Walker. ◗

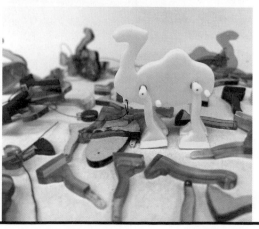

LOWER-COST
METAL 3D
PRINTING

Written by Joan Horvath and Rich Cameron

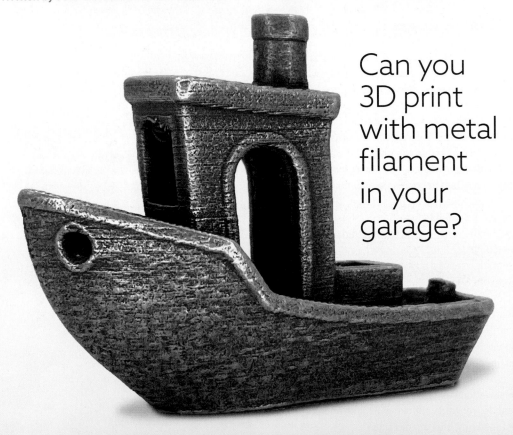

Can you 3D print with metal filament in your garage?

JOAN HORVATH and **RICH CAMERON** are the co-founders of Nonscriptum LLC (nonscriptum.com). They are the authors of many books, including *Make: Geometry* and *Make: Calculus*.

A

Metal 3D printing conjures up an image of lasers depositing energy into explosive powders, and systems that cost millions of dollars. However, there are viable options that will allow you to print metal parts on a consumer-level filament 3D printer. The investments start at a couple thousand dollars and go up a couple orders of magnitude from there. Let's find out what it takes to create metal 3D printed parts with garage-level maker tech.

LIKE MIM — BUT WITH FILAMENT

The roots of this technology lie in *metal-injection molding (MIM)*. MIM parts are created from a feedstock made of metal powder and a binder. The feedstock is typically 80% or higher metal by weight, but, since the metal is so much denser than the binder (maybe a factor of 7 or 8), the split by volume is closer to equal.

The mix is injection-molded to create a *green part.* Next is *debinding*: the binder is baked out or removed chemically to create what is called a *brown part.* Finally, the brown part is *sintered* at high temperature to form a metal part. After this process, the part will shrink 15% or so, typically more in the vertical direction due to gravity, and a bit less in the horizontal plane.

If we have an imaginary cube of the metal-binder mix that is 60% metal by volume (Figure **A**), the relative volumes of metal (gray) left after sintering and binder (blue) would be as shown. The ratio of the sides of these three cubes is 1.00: 0.84: 0.74.

Now imagine that instead of injection-molding the part, we made this metal-binder mixture into 3D printer filament (Figure **B**). The parallels between injection-molding a plastic part versus 3D printing a plastic part are pretty much the same for MIM versus extrusion 3D printing with high-metal-content filament. That is, the filament system avoids the need for creating molds, and makes it cost effective to make a single part. (Note that these high-metal filaments are different from filaments with less metal powder that create parts that only look like metal, but have properties closer to plastic.)

At a minimum, a system to 3D print high-metal-content filament consists of:

- a 3D printer
- a debinding method to remove the plastic binder — some companies rely on a chemical debinder, while others use heat
- a furnace to sinter the metal particles together.

MIM is a mature technology that has been around for a while, and those techniques (and

B

The Virtual Foundry, Rich Cameron

Green

Brown

Sintered

standards) gave a good jumpstart to companies trying to create consumer-level high-metal-content filament and metal extrusion printing workflows. What you need to try this at home varies, depending on which company's systems you decide to try out. Since binders and debinding and sintering processes are designed to work together, you'll need to buy into one end-to-end process or another.

High-metal-content filament is abrasive. Special *abrasion-resistant nozzles* made with materials like hardened steel, ruby, or sintered diamond are needed to print with these materials. A standard bronze nozzle will just be abraded away in very short order. That cost should be included in your budget for metal 3D printing. Some 3D printer manufacturers have add-on kits specifically for printing metal, with appropriate nozzle and sometimes bed modifications.

THE VIRTUAL FOUNDRY

The Virtual Foundry (thevirtualfoundry.com) focuses on users wanting to print metal parts at a minimum cost. Their filaments (branded Filamet) are designed for heat debinding, so a controllable furnace can do both steps. In addition to a hardened steel nozzle, the company also recommends a device to warm the filament (a Filawarmer) as it comes off the spool. This makes the filament more pliable and less likely to break.

Once you've set up your printer, you'll need

to print the part bigger in all three dimensions, by scaling it up in your slicing software or other means. Each material and process will require different scaling factors. The resulting green part (Figure **C**, left) can be sanded or otherwise cleaned up before the debinding process. A microscope image of a copper part surface at this green stage, showing how inhomogeneous the material is, can be seen in Figure **D**.

This green part is packed in a crucible with aluminum oxide Al_2O_3 (Figure **E**), which will support the print as the binder is baked out. The aluminum oxide is like beach sand and adds some stiffness to support the part. The crucible and kiln shown in Figure **F** are the ones listed in the price list later in this article. This kiln and can run on regular 120V wall power.

The result is the brown part (Figure C, center). Brown parts have had much of the binder removed, and can be fairly fragile. Brown parts in copper and bronze are about the consistency of a slightly stale store-bought cookie.

Finally the part is placed in a kiln, packed in talc to support it, heated to sinter together the metal, and then polished if desired (Figure C, right). Most metals also require carbon added to the talc to avoid oxidation, and other processes.

The finished part on the right was tumbled in a standard tumbler with stainless steel media, then polished with a brass wire wheel. The shrinkage (mostly during sintering) is clear in the series of images.

The Virtual Foundry

E

The Virtual Foundry's founder, Bradley Woods, demonstrates usage of the kiln.

F

Including a typical filament-based printer, the costs of a first part and second part are shown in **Table 1**. The amount of filament (in grams and meters), and resulting costs for the Benchy print shown in this article are: bronze 76g, 7.37m, $16.36; copper 83g, 7.37m, $12.99. The small crucible listed here (and shown in Figure F) can hold a part of diameter 43mm or less and height 58mm or less. Depending on when you read this, of course, prices can vary. You can wait for everything to cool to room temperature before touching it, but having additional safety equipment like forge gloves is never a bad idea.

Copper and bronze are the easiest metals to start with, since they have fewer oxidation issues during sintering. Steels require sintering carbon added to the sintering mix to absorb oxygen. Other materials sold by The Virtual Foundry are not really intended for desktop users, since they are reactive and need special handling, or hotter furnaces than the introductory one mentioned here. In some cases, users stop with a green part, just to make a part that is very heavy. In those cases, the part is just printed and used as-is, binder and all.

The final metal properties, such as how dense the final parts are, are acutely dependent on how carefully the debinding and sintering processes are managed. Density depends on the length of time the parts are sintered, among other factors.

The Virtual Foundry has extensive directions on their site for particular materials, and a Discord community for users to share experimental data and tips at thevirtualfoundry.com/tvfdiscord. If debinding and sintering feels like more than you want to deal with, their partner Sapphire3D

TABLE 1: FIRST PART COST

3D printer — Ender 3 S1	$322.00
Hardened steel nozzle, 0.6mm	$13.52
Filawarmer	$72.00
Print bed prep — blue painter's tape	$7.88
Sample Copper Filamet — 1.75mm	$20.59
Stainless steel crucible, 300ml	$7.63
Sintering refractory ballast — Al_2O_3 (1kg)	$44.00
Sintering refractory ballast — talc powder (0.5kg)	$35.04
Sintering carbon (0.5kg)	$44.00
Kiln paper	$19.39
Kiln — FireBox 8x4 LT Ceramic Kiln FK	$1,072.70
	$1,658.75

SECOND PART COST

Sample Copper Filamet — 1.75mm	**$20.59**

(sapphire3d.com) provides those steps as a service, with prices based on material and size of the part.

ZETAMIX

French company Nanoe (nanoe.com/en) produces a filament line called Zetamix (zetamix.fr/en), two versions of which are 316L and H13 steel. Their materials are more aimed at a small industrial user who wants to make tooling in-house with fast turnaround, like these parts printed in 316-L stainless steel (Figure **G** on the following page).

Nanoe separates their debinding and sintering steps. Chemical debinding occurs in an acetone bath. Sintering is done in a furnace that uses

gas like argon with a small amount of hydrogen, a common welding gas mixture. Their sintering furnace is designed to run at high temperatures with the part in this gas. As a result, this is a pricier system, with entry-level costs around $10,000.

In this process, brown parts retain some of their binder. Nanoe's COE, Guillaume de Calan, likes to make the analogy of building sand castles. Removing all of the binder would give you a pile of sand, but if some of the binder remains, the castle would still be damp and hold together. Sintering takes place in a gas environment, so supports may be needed through that stage.

Prices for 3D-printed parts are always very dependent on the details. Having said that, Nanoe gave us the example of a small lubrication nozzle that has to spray a very high-pressure fluid on machined parts. The part was just not possible to machine, due to its internal channel. It is approximately 5cm high and weighs 50g.

Cost for this print for a European customer would be around €25 ($27) of material, and around €120 per cycle run (€70 of electricity, argon gas, and consumables, and about €50 of furnace depreciation). You can put a few parts in a furnace run, so depending on the number of parts in the run, total cost per part would be lower — in the range of €55 for four parts per run, for example. Outsourced sintering of the same part was estimated at about €85. Sintering partners are listed on the Zetamix resellers page.

BASF ULTRAFUSE

If you don't want to wade into debinding and sintering yourself, chemical company BASF has created Ultrafuse 316L and 17-4 PH stainless steel filaments. The chemical debinding and sintering they require are industrial processes that are not feasible in a home environment. Because the sintering process takes place in a gas environment, supports would need to stay on through sintering for this option, and be removed with metalworking tools at the end of the process. (The provider just says parts cannot have overhangs for this reason.)

As with the other options in this article, at a minimum, you'll need a hardened steel nozzle for your printer. Once these modifications are made, though, at around $148 for a kilogram of filament plus $50 to sinter parts made from it, this might be the cheapest option. However, your control over the process is limited.

To debind and sinter the parts, users buy coupons (currently $50 for up to 1kg of parts) and ship them out to an industrial partner who runs a batch every couple of weeks. The maximum part size is 100mm on a side; other design constraints can be found on the Ultrafuse product pages at forward-am.com.

OTHER CONSIDERATIONS

The bottom line is: What system makes the most sense for you depends on how often you want to make a print; materials you are interested in; and your patience with tuning a process with a lot of variables. If you have a very short-term project, then outsourcing the debinding and sintering might make sense. If you have more challenging needs (and perhaps experience dealing with welding gases) you might want to invest in a more sophisticated furnace.

The filament is heavy: three kilograms of bronze filament from The Virtual Foundry is about the same length of filament as one kilogram of PLA. A 3kg spool might snap off your spool holder if it's flimsy.

Supports needed for printing are typically removed at the green part stage, when metalworking tools are not required. However, some thought might be required about the best orientation for sintering, which might not be the same as the orientation for printing. Be sure you understand the details of how the part is supported during the sintering process for the particular workflow you choose. Parts may warp

Nanoe, Desktop Metal, Rich Cameron

during sintering if this is not taken into account. If high dimensional accuracy is essential, you may need to step up to a more pricey system.

As with all 3D printing, slicing settings and preparation of the print bed are critical for good results, and you should consult the documentation for any manufacturer's product (and the rationale, which can be a little counterintuitive for those of us used to plastic).

NEXT STEPS

The next step up in sophistication and price is one of the larger systems that either uses some sort of feedstock with a binder mixed with powder, or sprays a binder on a powder. For example, Desktop Metal's Studio System (Figure **H**) uses metal feedstock rods instead of filament; their Shop System uses a jetted binder (see page 119). Markforged uses bound powder filament in their MetalX system. (In the interest of full disclosure, we did a curriculum-development job for Desktop Metal last year.)

These integrated systems come with software tools to make it easier and more predictable to create metal prints, but of course at higher cost than the more-DIY versions above. (See SJ Jones' starter guide on the following pages.)

ALTERNATIVES

A final question is whether your project actually needs to be printed in metal. There is a wide range of materials between PLA and stainless steel, and, depending on the real requirements, you might have better options.

If the part needs to be strong and stiff, a composite material (like continuous or chopped carbon fiber) might fit the bill. A composite part is likely to be a lot lighter than a metal one as well. If heat resistance is an issue, there are ceramic 3D-printable materials that are sintered similar to the process for metal. However ceramic debinding and sintering can be an easier process than the metal equivalents. ⊘

If you're interested in learning more, we have written a course on LinkedIn Learning. Check out "Additive Manufacturing: Metal 3D Printing" at linkedin.com/learning/additive-manufacturing-metal-3d-printing

SJ'S DECLASSIFIED
METAL PRINTING
STARTER GUIDE

Ready to try industrial metal 3D printing? Here's how the rocket scientists do it

Written by SJ Jones

SJ JONES, aka THEE Hottie of Metal Printing, is a metal additive solutions engineer who 3D prints metal parts to help solve critical supply chain challenges for aerospace and energy customers.

I started metal 3D printing back in 2018 for a small startup based on Florida's Space Coast. Our primary customers were large aerospace and industrial companies, and yes, we printed cool classified stuff I can't talk about. But we also saw plenty of startups looking to make prototypes, automotive hobbyists looking for replacement parts, university students looking to make the next big innovation, and regular folk who were merely curious about what we could (and could not) make. People love the idea of metal printing.

But most hobbyists feel boxed out of the metal additive industry because the printers come with a six- or seven-figure price tag. These are industrial pieces of capital equipment used to manufacture anything from satellites, tanks, airplanes, and supercars down to microchips and semiconductors. But don't let the sticker shock stop you from exploring what's possible! Consider this guide as your glance behind the curtain — at not just ways to get parts printed, but also at a possible career in industrial 3D printing.

TYPES OF INDUSTRIAL METAL 3D PRINTING

There are a variety of exotic approaches in high-end metal printing, but most systems are one of these two types:

Metal powder bed fusion (PBF) is a welding process wherein lasers — or sometimes electrons — hit tiny particles of metal powder and melt them together. And when you perform that melting layer by layer, it's roughly called metal 3D printing (Figure **A**).

PBF systems may also be called **direct metal laser sintering (DMLS), selective laser melting (SLM), or electron beam melting (EBM)** machines. Some familiar names are SLM, EOS, Velo3D, Stratasys, Freemelt, and Renishaw.

Metal binder jetting is a process of layer printing wherein a liquid binder is used to bond the metal powder rather than a laser. Then, those very fragile parts — in what is called a **green state** — will be **sintered** (or cured) in an oven to strengthen them (Figure **B**). The cured parts may be more porous and weaker than PBF-welded parts, but binder jetting is less expensive

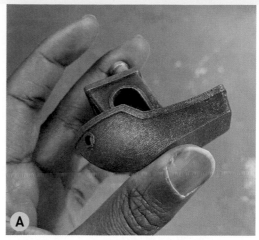

Benchy boat in nickel superalloy, printed without supports on a Velo3D laser powder bed fusion system.

The author working on an EOS Eosint M280 laser powder bed fusion 3D printer.

Complex part in 420SS/BR stainless steel/bronze, printed on the ExOne M-Flex metal binder jet system used by Sculpteo.

Siemens Energy, SJ Jones, Sculpteo

Customer installation of Desktop Metal's high-volume binder jet Production System.

A custom clutch plate printed on Desktop Metal's mid-volume binder jet Shop System, after sintering.

and suffers less warping. Examples are ExOne, Desktop Metal, Markforged, and HP.

HOW DO I GET PARTS MADE?

If you're a hobbyist looking to have prints made in metal, consider this a starter guide.

- **Material:** First and foremost, understand your material needs: Will a stainless steel or aluminum suffice? Or do you need something stronger like titanium or a nickel super alloy?
- **Size:** How big is the overall part? Size not only determines which printers can make your part, but it also has an influence on overall resolution.
- **Tolerance and surface:** How tight are your tolerances? If it's within 0.005 inches or less, then your part will most likely require machining post print. Resolution also plays an important role in surface finish. In the aerospace and energy industries, surface roughness can affect the way fuel flows through a part, for example, and can impact overall performance.

Once you have an understanding of the material and part size you need, that will narrow down the available printers. From there, **select your printer** based on your manufacturing requirements, tolerances, etc.

After you've found your printer, the next step is to get a clean and acceptable **file type**. Metal printing is slowly moving away from STLs; it's more and more common that solid CAD files or 3MF files are becoming the print standard.

The most common question I get about metal printing is, "Where can I go to have parts made?" The easiest option from my experience is to check out **Xometry**, which is an online manufacturing hub. There, you can get cost and lead time estimates up front based on the printer and material selected (Figures **C** and **D**). The offerings are limited to mostly aluminum and steel alloys, but it's still a fantastic starting point.

If you need a more expensive alloy such as nickel or copper, start by figuring out who has the printer you're looking for and then giving them a call. This second option will be quite frustrating as many manufacturers are reluctant to take in a one-off job that requires so much paperwork, but some can be persuaded with enough time and patience.

LIMITATIONS

For someone coming from the hobbyist world, it's important to keep in mind that metal printed parts can in some ways outperform conventionally made parts, but not without limitations. For example, binder jet parts in the green state are often weak and brittle, requiring more care in handling; they obtain their material strength through post process sintering, curing, and other finishing steps. Parts made with laser powder bed fusion tend to have rougher surfaces than machined parts.

The other limitation is consistency. If you order a large batch of metal printed parts, you may see some variation/warpage depending on which printer it was made on, especially if the parts are

Desktop Metal, Ellie Rose

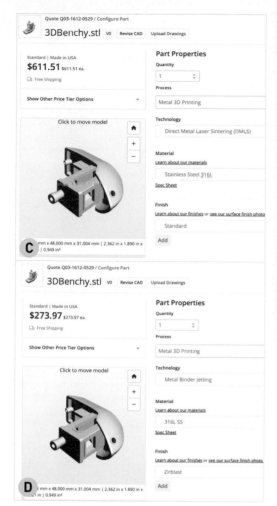

Metal extrusion part in green state, pre-sintering, made with BASF Ultrafuse 316L steel filament, courtesy of Ellie Rose (@elianarose66).

Price quotes for a 60mm (2.36") 316L stainless steel Benchy boat printed on laser PBF ($600+) or binder jet ($270+) metal printers. The same job in ABS plastic would cost about $24.

removed from the plates with no post processing in order to save on cost.

CHALLENGES IN METAL 3D PRINTING

The biggest challenge we have in metal 3D printing right now is qualifying parts for actual use, especially if that use involves human lives. We have to certify that all the parts we've printed are sound, safe, and proven to perform in the most extreme operating conditions. In many cases, that means we need to be better than conventionally made parts. But the industry is currently in deep discussion over which data in

the manufacturing process is most important for this qualification — is it temperature, powder quality, oxygen content, all of the above?

The second biggest challenge we have is scaling our operations. In metal printing, the printing aspect is a small part of what goes into making a part. There are thermal treatments that occur after printing to adjust the material properties, machining to rein in tight tolerances, surface finishing to get a smooth polish, and so on. These additional steps are easy when we're talking batches of two or four or eight parts, but on the grand scale of hundreds it can be quite costly and turn into a logistical nightmare if not managed properly.

Our third largest challenge is workforce. We need more makers — more people who think outside the conventional way things are done. ◢

Watch: SJ's Declassified Metal Survival Guide, from TIPE 3D Printing 2021 conference by Women in 3D Printing: youtu.be/Y6qIeI4hNCl

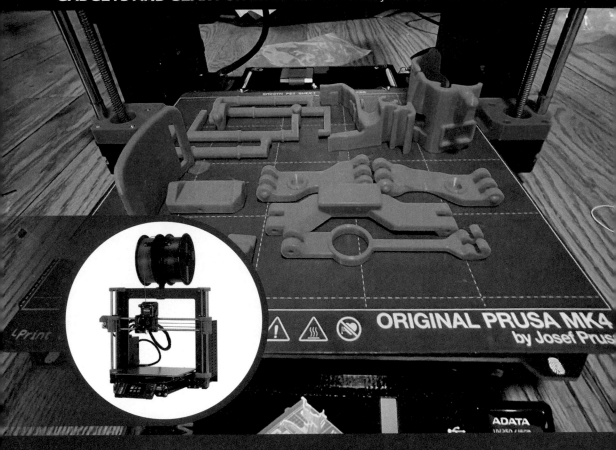

Prusa MK4 3D Printer $1,099 prusa3d.com

The latest update to the Prusa MK series of printers has been in the works for many years. The MK3 which came out in 2017 has been considered the industry standard to meet when it comes to quality and reliability.

The MK4 continues the tradition of building on the predecessors with an upgrade path for previous owners. This is awesome, in that folks can upgrade their old machines, but it also leaves the MK4 feeling a bit dated, at least visually.

Featurewise, the upgrades may not be headline worthy on their own; there's a load cell, a new extruder, a few tweaks to the frame, and a new color screen. However, when it comes to actually using the machine, the upgrades can't be overlooked.

The load cell is there to ensure a perfect first layer and theoretically detect crashes and errors.

It works really well too — my first layers have been absolutely beautiful.

With the latest competition being speed, Prusa advertised an updated firmware "coming soon" to give this machine a more competitive speed boost. At the time of writing, I haven't tested that.

I have been putting this machine through hours and hours of demanding print jobs and I've been very happy with the results. The super reliable first layer gives much-wanted peace of mind. I also appreciate that I'm supporting an open source project that embraces the 3D printing community and their contributions.

Ultimately, if this machine is as reliable as my MK3S but with a perfect first layer all the time, I'm happy about it. My MK3S is still, even after years of aging, the most reliable printer I've ever used. —Caleb Kraft

Voxelab Aquila D1

$399 voxelab3dp.com

The D1 feels super sturdy. The metal frame and metal brackets all make it just feel solid. It doesn't wiggle or creak at all, not even the screen.

Firing up the machine, there is a constant fan noise from the power supply. This isn't horrible, but it is somewhat noisy and you'll hear it constantly.

Print quality is fantastic. I see no z-banding or layer issues at all. The textured sheet means that those first layers are going to be just beautiful, if you like that textured look.

Overall, this is a very usable machine. It seems to be printing quite well in terms of quality and reliability. However, there are a few quirks I found that I'd love to see addressed in future updates to this line.

The biggest issue is the cable management. The cable that goes to the extruder just flops around. There's no way to affix it anywhere. The point where it attaches to the extruder actually impedes the filament path. I found this surprising; anyone assembling this printer and using it even once would see this as an issue.

At the price, this thing seems to be packed with features. It isn't blazing fast, but so far, it has produce great quality without any headaches. —*Caleb Kraft*

Mingda Magician X2

$269 3dmingdaofficial.com

Like the last Magician model, assembling the X2 is super quick and easy, only a few bolts. You plug the corresponding wires into the motors, extruder, and bed and you're ready to go.

I really like how they've done their cable management. They use very flat ribbon cables that line up well with the axis of motion. They never feel in the way or at risk of pinching.

The first thing I did was level the bed, which went smoothly. Then I loaded some filament and selected a pre-prepared print from the card and hit Print. I got a perfect little bunny! The quality is fantastic, even in the tall thin ears, which might have been nasty if there wasn't enough cooling.

On the website they call out their LED temperature indicator on the extruder. I don't know how useful it is, but it's pretty.

The machine is fairly quiet in operation, which is nice. The Mingda interface feels a bit dated, but it does the job just fine. Labels like "percentage" don't really tell you much, but with some poking around you'll figure out what it all means.

Overall, you're getting good quality at the price. It seems Mingda has made a pretty solid machine for under $300. —*Caleb Kraft*

Adafruit Metro M7 With AirLift

$30 adafruit.com/product/4950

The latest Adafruit Metro is the M7 with AirLift. Adafruit have essentially taken a very powerful ARM Cortex-M7 microcontroller, put it in the Arduino Uno's form factor, and then squeezed in a Wi-Fi subsystem along the edge, plus all the extras found on their recent boards.

The Metro M7 is built around the NXP i.MX RT1011, which is an Arm Cortex-M7 processor running at 500MHz. NXP calls this a "crossover" microcontroller because it aims to have all the power of an A-class application processor and the real-time control capabilities of an M-class microcontroller in a single core. The chip has 128kB of RAM, a floating point unit, and digital signal processing built in. It uses external flash over a QSPI bus for program storage, with Adafruit providing 4MB on the board.

The other microcontroller on the board is the ESP32-based AirLift Wi-Fi co-processor, connected to the main microcontroller via SPI and a CircuitPython library written by Adafruit. Extras on the board include status LEDs, a NeoPixel, an on/off switch, and a Stemma QT connector. Overall the new Metro M7 offers greatly increased processing power, with Wi-Fi offloaded to a second microcontroller.

— *Paul J. Henley*

HTVRONT Auto Heat Press

$251 bit.ly/3lAyqgl

There is basically no setup with this machine. You unbox it and it is ready to go. It has a few presets that might fit the jobs you need, but you can always adjust things manually.

A normal "clamshell" heat press opens up, you guessed it, like a clam. This exposes the hot surfaces to your hands and typically these take a fair amount of pressure to close. The HTVRONT automatic system is built differently. The garment goes on a slide-out tray that keeps your hands away from the heated platen. You don't have to use force at all, as the system compresses things at the push of a button.

The price feels very competitive to me, as the last programmable clamshell I purchased was basically the same price. If you lack upper body strength or just want a nice machine where you can program a few preset modes, this thing is really nice.

My only concern is that you can see the machine actually flex as it applies pressure to the garment. I have no idea if this is intentional as part of the design — possibly the "smart pressure transducer" — or if the frame is just flexing.

I really like this machine. The display makes sense, the automatic pressing is a very nice comfort feature, and the whole thing just feels really solid. —*Caleb Kraft*

Much Ado About Almost Nothing by Hans Camenzind

$23 amazon.com/dp/1949267962

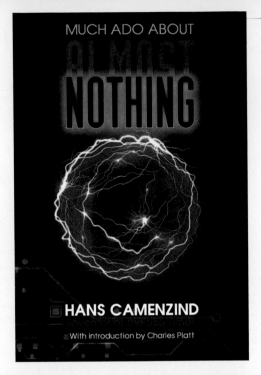

You may recognize Hans Camenzind's name as the inventor of the 555 timer chip. What you may not know is that he was a uniquely talented writer.

Five years before his death in 2012, he published *Much Ado About Almost Nothing*, which I regard as the most fascinating, revelatory, and enlightening history of electricity and electronics ever written. Based on a huge amount of research, it documents the ambitions, struggles, successes and (in some cases) tragedies of brilliant men who used the humble electron to revolutionize the world we live in.

Before I read this book, I never knew that it took more than 50 years for Benjamin Franklin's invention of the lightning rod to be adopted in Europe, because people were afraid that it *attracted* lightning. I was unaware that Alexander Graham Bell initially wanted to construct a telephone using reeds that vibrated like tuning forks. I didn't know that Tesla's biggest generator, in Colorado Springs, used a coil 100 feet in diameter which disturbed horses in the neighborhood by inducing voltages in their hooves.

The book is crammed with such fascinating details, and also describes the epic patent wars between inventors and huge corporations such as RCA that tried to steal their work. At the same time, Camenzind digresses into lucid explanations of the phenomena and techniques that made electronics possible, culminating in the planar process of chip making, conceived by Jean Hoerni, who remains relatively obscure even though his idea was fundamental to the semiconductor industry.

New York publishers apparently were uninterested in this book, perhaps because editors couldn't categorize it. What *was* it, anyway? A series of sometimes scandalous portraits of obsessional inventors? A slightly technical explanation of electricity and electronics? Well, it was both! But in that case, bookstores wouldn't know where to put it. And why did it have a title that no one would understand?

In the end, it was published by a small company that did not promote it. *Much Ado* became one of the best-kept secrets in book publishing. I came across it while writing my tribute to Hans Camenzind for *Make:* Volume 82. When I asked his widow, Pia, if she would permit a new edition, she was open to that idea, so I placed it with Faraday Press. When asked if I would write an introduction, I was happy to oblige.

I have no financial interest in all this. I just wanted to give a great book a new lease on life. I don't seriously believe it can be a best-seller, even though it deserves to be. What I do believe is that if you have any interest in electronics, you will find it compulsive reading.

—*Charles Platt*